ITSM: An Interactive Time Series Modelling Package for the PC

Peter J. Brockwell   Richard A. Davis

ITSM:
An Interactive Time Series
Modelling Package for the PC

With 53 Illustrations and 3 Diskettes

Written in collaboration with R.J. Hyndman

Springer-Verlag
New York Berlin Heidelberg London
Paris Tokyo Hong Kong Barcelona

Peter J. Brockwell
Richard A. Davis

Department of Statistics
Colorado State University
Fort Collins, CO 80523
USA

Library of Congress Cataloging-in-Publication Data
Brockwell, Peter J.
    ITSM: an interactive time series modelling package for the PC/
Peter J. Brockwell, Richard A. Davis.
    p. cm.
    Includes index.
    ISBN-13:978-0-387-97482-8
    1. ITSM (Computer file) 2. Time-series analysis—Data processing.
I. Davis, Richard A.   II. Title.
QA280.B757 1991
519.5'5'0285—dc20   90-19622

Photocomposed copy prepared using LaTex.

9 8 7 6 5 4 3 2 1

ISBN-13:978-0-387-97482-8   e-ISBN-13:978-1-4612-3116-5
DOI: 10.1007/978-1-4612-3116-5

Preface

The package **ITSM** (Interactive Time Series Modelling) evolved from the programs for the IBM PC written to accompany our book, *Time Series: Theory and Methods*, published by Springer-Verlag. It owes its existence to the many suggestions for improvements received from users of the earlier programs.

The requirements for running the programs are as follows:

- IBM PC, PC/XT, PC/AT or compatible computer operating under MS-DOS;

- at least 540 K of RAM available for applications;

- a CGA, EGA, VGA or Hercules card for graphics;

- a mathematics co-processor (recommended but not essential).

The package includes the screen editor *WORD6* and six programs, *PEST*, *SMOOTH*, *SPEC*, *TRANS*, *ARVEC* and *ARAR*, whose functions are summarized in Chapter 1.

Two of the programs, *ARVEC* (for multivariate autoregressive modelling) and *ARAR* (for univariate forecasting), have been added to the programs contained in earlier versions of the package. The other programs have been corrected and improved in many respects and the use of graphics has been considerably expanded.

The programs *PEST*, *SPEC* and *SMOOTH* can be used to analyze time series of up to 2300 observations; *ARVEC*, *ARAR* and *TRANS* can handle series of lengths up to 1000, 1000 and 800 respectively.

We are greatly indebted to many people associated with the development of the programs and manual. Outstanding contributions were made by Joe Mandarino, the architect of the original version of *PEST*, Rob Hyndman, who wrote the original version of the manual for *PEST*, and Anthony Brockwell, who provided *WORD6*, the graphics subroutines and general computing expertise. The first version of the *PEST* manual was prepared for use in a short course given by the Key Centre in Statistical Sciences at RMIT and The University of Melbourne. We are indebted to the Key Centre for support and for permission to make use of that material. We also wish to thank the National Science Foundation for support of the research on which many of the algorithms are based, R. Schnabel of the University of Colorado computer science department for permission to use his optimization program, and Carolyn Cook for her assistance in the final preparation of the manual. We are grateful for the encouragement provided by Duane Boes and the excellent working environments of Colorado State University and the University of Melbourne. The editors of Springer-Verlag have been a constant source of support and encouragement and our families, as always, have played a key role in maintaining our sanity.

Fort Collins, Colorado   PETER J. BROCKWELL
August, 1990   RICHARD A. DAVIS

Contents

1

Introduction

1.1 The Programs

The time series programs described in this manual are all included in the package **ITSM** (Interactive Time Series Modelling) designed to accompany the book *Time Series: Theory and Methods* by Peter Brockwell and Richard Davis, (Springer-Verlag, Second Edition, 1991). With this manual you will find a two-diskette $5\frac{1}{4}$" version as well a single-diskette $3\frac{1}{2}$" version of the package. References in the manual to Disk 1 and Disk 2 apply only if you are using the $5\frac{1}{4}$" version.

PEST is a program for the modelling, analysis and forecasting of univariate time series. The name "PEST" is an abbreviation for **P**arameter **EST**imation.

SPEC is a program which performs non-parametric spectral estimation for both univariate and bivariate time series.

SMOOTH permits the user to apply symmetric moving average or one-sided exponential smoothing operators to a given data set.

TRANS allows the calculation and plotting of sample cross-correlations between two series of equal lengths, and the fitting of a transfer function model to represent the relation between them.

ARVEC fits vector autoregressive models to multivariate time series with up to 6 components and allows automatic order-selection using the AICC criterion.

ARAR is based on the ARARMA forecasting technique of Newton and Parzen. For a univariate data set it first selects and applies (if necessary) a memory-shortening transformation to the data. It then fits a subset autoregressive model to the memory-shortened series and uses the fitted model to calculate forecasts.

This manual is designed to be a practical guide to the use of the programs. For a more extensive discussion of time series modelling and the methods used in **ITSM**, see the book *Time Series: Theory and Methods*, referred to subsequently as *BD* . Information regarding the data sets on Disks 1 and 2 is contained in Appendix B. Further details, and in some cases analysis of the data, can be found in *BD* .

1.2   System Requirements

- IBM PC, PC/XT, PC/AT or compatible computer operating under MS-DOS;

- at least 540 K of RAM available for applications;

- a CGA, EGA, VGA or Hercules card for graphics;

- a mathematics co-processor (recommended but not essential).

When booting the computer, the program ANSI.SYS should be loaded. This is done by including the command DEVICES=ANSI.SYS in your CONFIG.SYS file.

Each of the programs is run through a batch file, e.g. PEST.BAT. To run this program, simply place Disk 1 in Drive A and type A: ↩ . Then type PEST↩ . (↩ denotes the <Enter> or <Return> key.) You will next be asked to enter a graphics code number (1,2,3 or 4) to indicate the type of graphics adaptor which you are using. Normally you will enter 1 for Hercules, 2 for CGA, 3 for EGA or VGA and 4 for High Resolution NCR Graphics. If you type the correct number followed by ↩ , you will see a four-line title enclosed in a rectangular box with a two-line message immediately below the box (see Figure 2.1). If you do not see the rectangle (or if your screen goes blank) type 1↩ . If this does not work type 2↩ , then 3↩ , then 4↩ , continuing until you get the required display. Make a note of your correct code number since it is used in all of the time series programs. Most EGA/VGA cards will work either with graphics code 2 or 3 but choice 3 will give higher resolution.

To obtain printed copies of graphs and text displayed on the screen, it is necessary to load a graphics dump program. The batch files, PEST.BAT, SPEC.BAT etc., automatically load the program *HIGHRES*, a graphics dump program for Epson and IBM dot-matrix printers. Graphs and text which subsequently appear on the screen can then be printed by holding down the <Shift> key and pressing the <Prt-Scr> key. To obtain a continuous record of the text printed on the screen, hold down the <Control> key and press <Prt-Scr> to begin printing. Holding down the <Control> key and pressing <Prt-Scr> a second time terminates printing.

1.2.1   HARD DISK INSTALLATION

Although the programs can be run directly from the disks provided, care is needed to ensure that you do not run out of storage space for filing results and new data. It is therefore recommended that the programs be installed on a hard disk.

To install **ITSM** on the hard disk C, place Disk 1 in Drive A, type
A:install A C

and follow the prompts printed on the screen. A directory called ITSM with subdirectory ITSM\DATA will then be created on Drive C and all the required files transferred. Other drives may be substituted for A and C in the install command depending on your system configuration.

Throughout this manual we shall assume that you have either installed the programs as described with ITSM as your current directory or that you are operating from one of the disks provided using either Drive A or Drive B.

1.3   Creating Data Files

All data to be used in the programs (except those for *ARVEC*) should be stored in standard ASCII files in column form. That is, each value must be on a separate row. The programs will read the first item of data from each row. Most of the data sets used in *BD* (and a number of others) are included on the diskettes in this form. Data sets for *ARVEC* are multivariate, with the m components observed at time t stored in row t of the file. (See for example the 150 observations of the bivariate series contained in the file LS.2 in the subdirectory DATA of Diskette 2.)

All data files can be examined and edited using *WORD6* — the screen editor provided on the diskette. New data files can also be created using *WORD6*. For example, to create a data file containing the numbers 1 to 5:

- Type WORD6↩ to call the screen editor *WORD6*.

- Then type
$$1↩ 2↩ 3↩ 4↩ 5↩$$

- Hold down the <Alt> key and press W. You will be asked for a filename. Type TEST↩ . Your new data file consisting of the column of numbers 1 2 3 4 5 will then be stored on your disk under the name TEST.

- To leave *WORD6*, hold down the <Alt> key again and press X.

- To check that your new file is on the diskette, type dir↩ . You should see a file called TEST on the listed contents of your directory.

- To read your new file, call *WORD6* again by typing WORD6↩ . Then hold down the <Alt> key and press R. You will be asked for a file name. Type TEST↩ . Your new data file consisting of the column of numbers 1 2 3 4 5 will then be read into *WORD6* and printed on the screen. Alternatively, you may read your new file into *WORD6* by typing WORD6 TEST↩ .

For further information on the use of *WORD6* see Appendix A.

2

PEST

2.1 Getting Started

2.1.1 RUNNING PEST

Simply type PEST↩ . You will then be asked to type 1↩ , 2↩ , 3↩ or
4↩ to indicate whether your computer has a Hercules, CGA, EGA/VGA
or NCR High-resolution graphics adaptor. After typing the appropriate
response you should see the screen displayed in Figure 2.1. If you do not
see the rectangle (or if your screen goes blank) type 1↩ , then 2↩ , then
3↩ , then 4↩ until you obtain the required display. (If you are unable to
get the required display, check that ANSI.SYS has been loaded as described
in Section 1.2.) Make a note of your graphics code number as it is used for
all the programs with graphics. Most EGA/VGA cards will work either
with graphics code 2 or 3 but choice 3 will give higher resolution.

Once you have obtained the display in Figure 2.1 type 0↩ to continue
and you will see the Main Menu of *PEST* as shown in Figure 2.2.

As you can see, only a few options are available. Other options will appear
in the Main Menu after further information (such as data) is entered.

PEST is menu-driven so that you are required only to make choices between options specified by the program. For example, if you choose Option
1 of the Main Menu (Data entry; statistics; transformations) by typing 1↩ ,
you will see the Data Menu, from which you can make a further selection,
e.g. Input new data set. To return to the Main Menu type 11↩ .

There are several distinct functions of the program *PEST*. The first is
to plot, analyze and possibly transform **time series data**, the second is
to compute properties of **time series models**, and the third utilizes the
previous two in **fitting models to data**. The latter includes checking that
the properties of the fitted model match those of the data in a suitable sense.
Having found an appropriate model, we can (for example) then use it in
conjunction with the data to forecast future values of the series. Sections
2.2-2.5 and 2.7 of this manual deal with the modelling and analysis of data,
while Section 2.6 is concerned with model properties.

It is important to keep in mind the distinction between data and model
properties and not to confuse the data with the model. At any particular
time *PEST* typically stores one data set and one model (which can be
identified using Option 10 of the Main Menu). Rarely (if ever) is a real time
series generated by a model as simple as those used for fitting purposes.
Our aim is to develop a model which mimics important features of the data,
but is still simple enough to be used with relative ease.

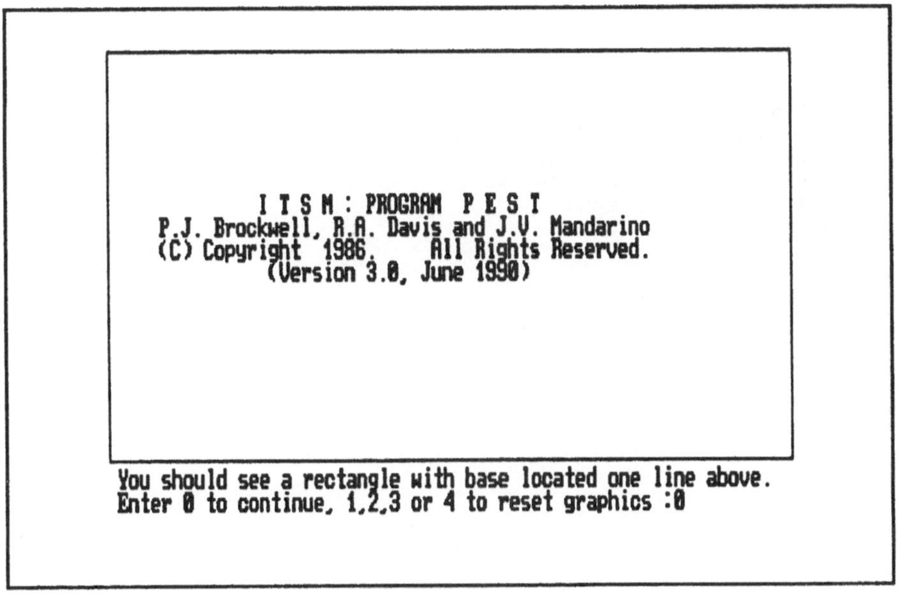

```
            I T S M : PROGRAM  P E S T
 P.J. Brockwell, R.A. Davis and J.V. Mandarino
 (C) Copyright  1986.      All Rights Reserved.
            (Version 3.0, June 1990)
```

```
You should see a rectangle with base located one line above.
Enter 0 to continue, 1,2,3 or 4 to reset graphics :0
```

FIGURE 2.1. *What you should see after entering the correct graphics number*

2.1.2   PEST TUTORIAL

The EXAMPLES in this chapter constitute a tutorial session for *PEST* in serialized form. They lead you through a complete analysis of the well-known Airline Passenger Series of Box and Jenkins (see Appendix B).

2.2   Preparing Your Data for Modelling

Once the observed values of your time series are available in a single-column ASCII file (see Section 1.3), you can begin model fitting with *PEST*. The program will read your data from the file, plot it on the screen, compute sample statistics and allow you to do a number of transformations designed to make your transformed data representable as a realization of a zero-mean stationary process.

EXAMPLE: To illustrate the analysis we shall use the data file AIRPASS, which contains the number of international airline passengers (in thousands) each month from Jan '49 to Dec '60.

```
MAIN MENU :
   1: Data entry; statistics; transformations
   2: Entry of an ARMA(p,q) model
   10: Model and data file status
   11: Exit from program
CHOOSE A NUMBER : 1
```

FIGURE 2.2. *The Main Menu of PEST*

2.2.1 ENTERING DATA

From the Main Menu of *PEST* select Option 1 (Data entry; statistics; trans-formations) by typing 1↩ . The Data Menu will then appear. Again choose Option 1 and you will be asked to confirm that you wish to enter new data. Respond by typing y and you will be asked for the name of the new data file. Type the name of the file and *PEST* will read your data.

A new data file can always be imported using Option 1 of the Data Menu. Note however that the previous data file is eliminated from *PEST* each time a new file is read in.

> EXAMPLE: Go through the above steps to read the airline pas-senger data into *PEST*. The file name is AIRPASS. Once the file has been read in, the screen should appear as in Figure 2.3.

2.2.2 FILING DATA

You may wish to change your data using *PEST* and then store it in another file. At any time after transforming the data in *PEST*, the transformed data

```
NAME OF THE DATAFILE TO BE ANALYZED : airpass

   TOTAL OBSERVATIONS =    144

   CHECKING THE FIRST THREE AND LAST DATA POINTS:

      112.0000000        118.0000000        132.0000000        432.0000000

   DATA FILE = airpass

DATA MENU :
    1. Input new data set
    2. Plot the  144 data values; find mean and variance
    3. Plot sample ACF/PACF of current data file
    4. File sample ACF/PACF of current data file
    5. Box-Cox transformation    [NOT after 6,7,8 or 9]
FOR CLASSICAL DECOMPOSITION USE 6 AND/OR 7. FOR DIFFERENCING USE 8.
    6. Remove seasonal compt.    [NOT after 7,8 or 9]
    7. Remove polynomial trend   [NOT after 8 or 9]
    8. Difference current data    [NOT after 6,7,or 9]
    9. Subtract the mean of             280.29860
    11. Return to main menu
CHOOSE A NUMBER :
```

FIGURE 2.3. *The PEST screen after reading in the AIRPASS data*

can be filed by choosing Option 10 from the Data Menu. Do not use the name of a file that already exists or it will be overwritten. When new data is first read into *PEST*, Option 10 will not be visible.

Whenever you file the current data after making only a Box-Cox transformation (see Section 2.2.4), the data stored in *PEST* will be renamed to match the new file name. (*PEST* also remembers the transformation so that it can later be inverted.)

2.2.3 PLOTTING DATA

The first step in the analysis of any time series is to plot the data. With *PEST* the data can be plotted by selecting Option 2 from the Data Menu. This will first produce a histogram of the data; pressing any key then causes a graph of the data vs. time to appear on the screen.

Under the histogram several sample statistics are printed. These are defined as follows:

Mean:

$$\bar{X} = \frac{1}{n}\sum_{i=1}^{n} X_i$$

Standard Deviation:

$$s = \sqrt{\frac{1}{n}\sum_{i=1}^{n}(X_i - \bar{X})^2} = \sqrt{\frac{1}{n}\left(\sum_{i=1}^{n}X_i^2 - n\bar{X}^2\right)}$$

Coefficient of Skewness:

$$\hat{\nu}_3 = \sqrt{\frac{1}{n}\sum_{i=1}^{n}(X_i - \bar{X})^3} = \sqrt{\frac{1}{n}\left(\sum_{i=1}^{n}X_i^3 - 3\bar{X}\sum_{i=1}^{n}X_i^2 + 2n\bar{X}^3\right)}$$

EXAMPLE: Continuing with our analysis of the data file AIR-PASS, choose Option 2 from the Data Menu. The first graph displayed is a histogram of the data, shown in Figure 2.4. Then press any key to obtain the time-plot shown in Figure 2.5. Finally press any key and type 0↩ to return to the Data Menu.

2.2.4   TRANSFORMING DATA (*BD Sections 1.4, 9.2*)

Transformations are applied in order to produce data which can be successfully modelled as *stationary time series*. In particular we need to eliminate trend and cyclic components and to achieve approximate constancy of level and variability with time.

EXAMPLE: The airline passenger data are clearly not stationary. The level and variability both increase with time and there appears to be a large seasonal component (with period 12).

Non-stationary data must be transformed before attempting to fit a stationary model. *PEST* provides a number of transformations which are useful for this purpose.

BOX-COX TRANSFORMATIONS (*BD Section 9.2*)

Box-Cox transformations can be carried out by selecting Option 5 of the Data Menu. If the original observations are $Y_1, Y_2, \ldots, Y_n$, the Box-Cox transformation f_λ converts them to $f_\lambda(Y_1), f_\lambda(Y_2), \ldots, f_\lambda(Y_n)$, where

$$f(y) = \begin{cases} \frac{y^\lambda - 1}{\lambda}, & \lambda \neq 0, \\ \log(y), & \lambda = 0. \end{cases}$$

These transformations are useful when the variability of the data increases or decreases with the level. By suitable choice of λ, the variability

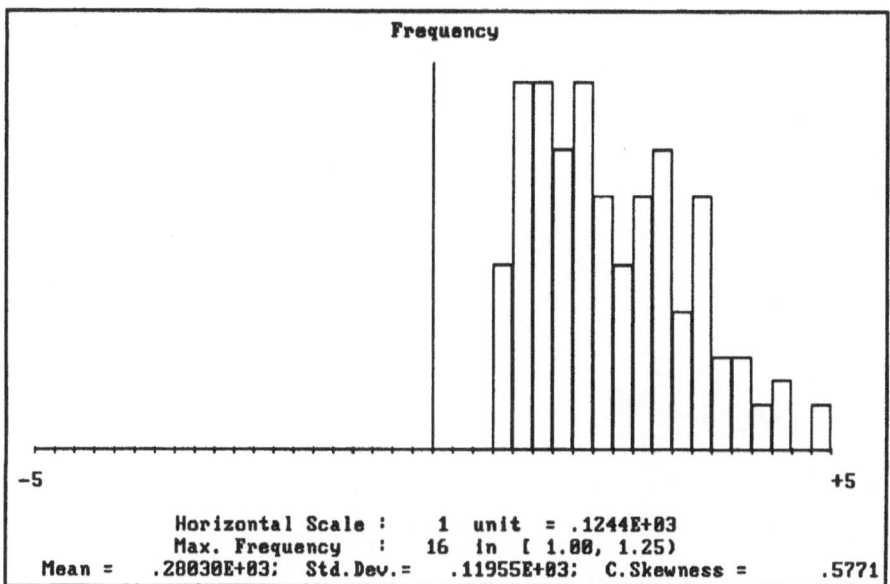

FIGURE 2.4. *The histogram of the AIRPASS data*

can often be made nearly constant. In particular, for positive data whose standard deviation increases linearly with level, the variability can be stabilized by choosing $\lambda = 0$.

The choice of λ can be made by trial and error, using the graphs of the transformed data which can be plotted using Option 2 of the Data Menu. (After inspecting the graph for a particular λ you can invert the transformation using Option 5 of the Data Menu, after which you can then try another value of λ.) Very often it is found that no transformation is needed or that the choice $\lambda = 0$ is satisfactory.

> EXAMPLE: For the AIRPASS data the variability increases with level and the data are strictly positive. Taking natural logarithms (i.e. choosing a Box-Cox transformation with $\lambda = 0$) gives the transformed data shown in Figure 2.6. You can perform this transformation and plot the graph (starting in the Data Menu with the data file AIRPASS) by typing 5↩ 0↩ (to transform the data), then 2↩ ↩ (to plot the graph).
>
> Notice how the variation no longer increases. The seasonal ef-

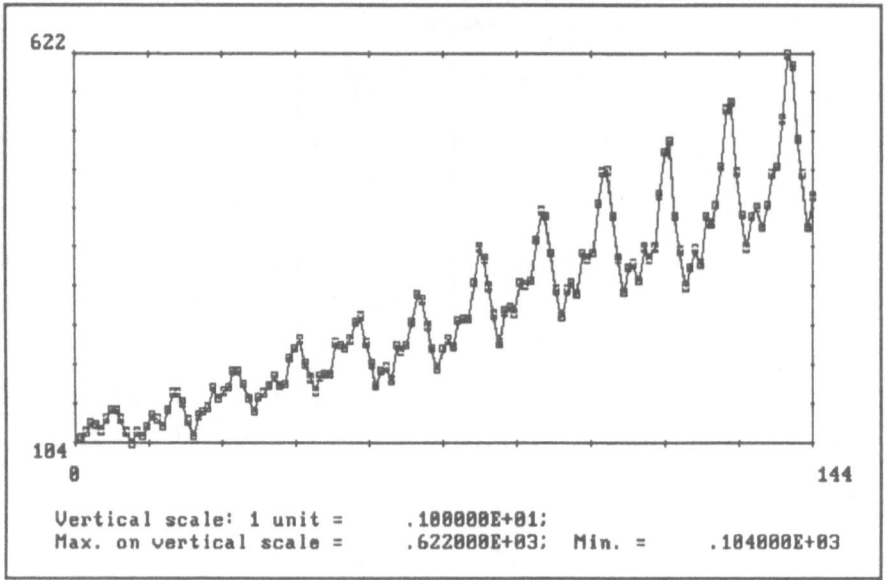

FIGURE 2.5. *The time-plot of the AIRPASS data*

fect remains, as does the upward trend. These will be removed
shortly. Since the log transformation has stabilized the variability, it is not necessary to consider other values of λ. Note that
the data stored in *PEST* now consists of the natural logarithms
of the original data.

Later it will be useful to have these "logged" data stored in a
file. Do this by choosing Option 10 from the Data Menu and
entering the file name AIRPASS.LOG. Observe that after filing,
the name of the data file stored in *PEST* is changed to match
the new file name AIRPASS.LOG (see Section 2.2.2).

CLASSICAL DECOMPOSITION (*BD Section 1.4*)

There are two methods provided in *PEST* for the elimination of trend and
seasonality. These are

(i) "classical decomposition" of the series into a trend component, a seasonal component and a random residual component and

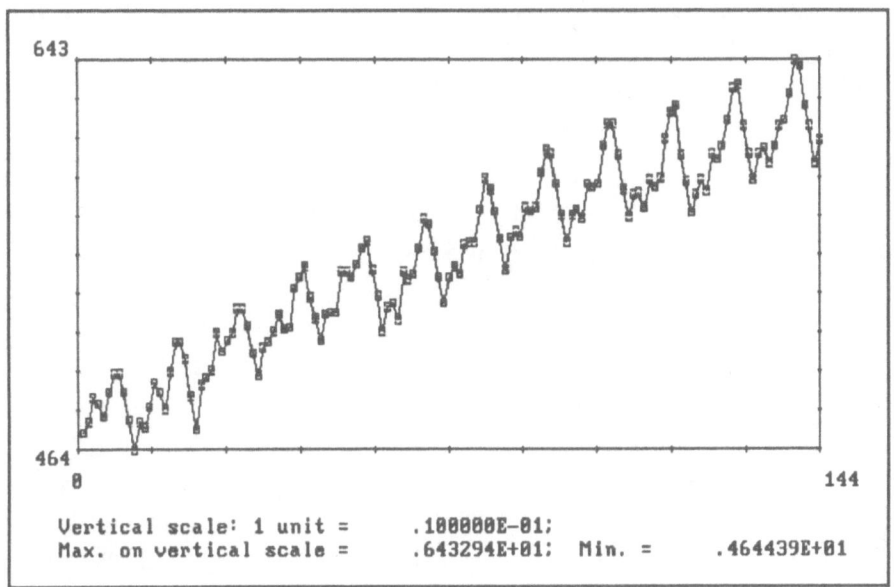

643

464

0   144

Vertical scale: 1 unit =   .100000E-01;
Max. on vertical scale =   .643294E+01;   Min. =   .464439E+01

FIGURE 2.6. *The AIRPASS data after taking logs*

(ii) differencing.

Classical decomposition of the series $\{X_t\}$ is based on the model,

$$X_t = m_t + s_t + Y_t$$

where X_t is the observation at time t, m_t is a "trend component", s_t is a "seasonal component" and Y_t is a "random noise component" which is stationary with mean zero. The objective is to estimate the components m_t and s_t and subtract them from the data to generate a sequence of residuals (or estimated noise) which can then be modelled as a stationary time series.

To achieve this, select Option 6 then Option 7 from the Data Menu. (You can also estimate trend only or seasonal component only by selecting the appropriate option separately.)

The estimated noise sequence automatically replaces the previous data stored in *PEST*.

EXAMPLE: The logged airline passenger data has an apparent seasonal component of period 12 (corresponding to the month

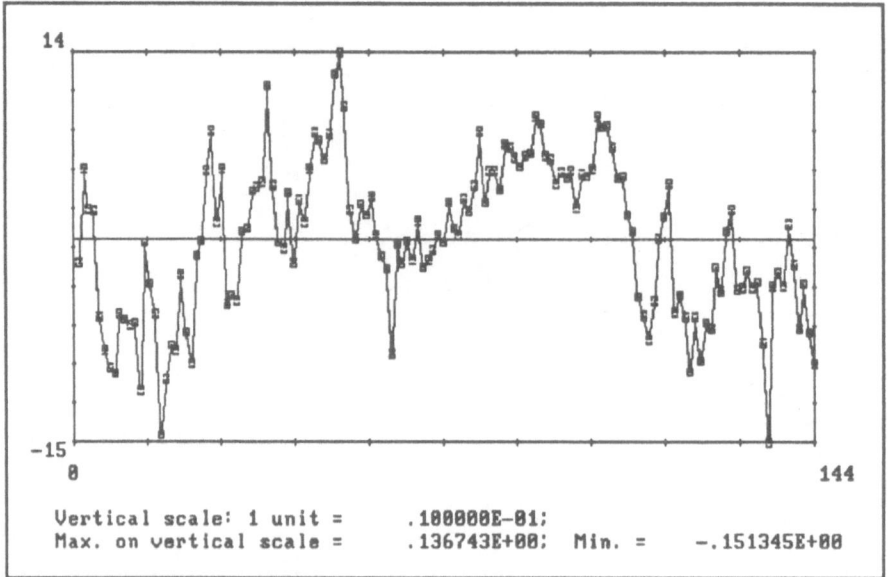

Vertical scale: 1 unit =   .100000E-01;
Max. on vertical scale =   .136743E+00; Min. =   -.151345E+00

FIGURE 2.7. *The series AIRPASS.LOG after classical decomposition*

of the year) and an approximately linear trend. Remove these
by typing 6↩ 12↩ ↩ 7↩ 1↩ ↩ (starting from the Data
Menu).

Figure 2.7 shows the transformed data (or residuals) Y_t, obtained by classical decomposition of the series AIRPASS.LOG.
$\{Y_t\}$ shows no obvious deviations from stationarity and it would
now be reasonable to attempt to fit a stationary time series
model to this series. We shall not pursue this approach any
further in our tutorial, but turn instead to the **differencing**
approach. (After completing the tutorial, you should have no
difficulty in returning to this point and completing the classical
decomposition analysis by fitting a stationary time series model
to $\{Y_t\}$.)

Restore the original airline passenger data into *PEST* by using
Option 1 of the Data Menu and reading in the file AIRPASS.

DIFFERENCING (*BD Sections 1.4, 9.1, 9.6*)

Differencing is a technique which can also be used to remove seasonal components and trends. The idea is simply to consider the differences between pairs of observations with appropriate time-separations. For example, to remove a seasonal component of period 12 from the series $\{X_t\}$, we generate the transformed series,

$$Y_t = X_t - X_{t-12}.$$

It is clear that all seasonal components of period 12 are eliminated by this transformation, which is called **differencing at lag 12**. A linear trend can be eliminated by differencing at lag 1, and a quadratic trend by differencing twice at lag 1 (i.e. differencing once to get a new series, then differencing the new series to get a second new series). Higher-order polynomials can be eliminated analogously. It is worth noting that differencing at lag 12 not only eliminates seasonal components with period 12 but also any linear trend.

Repeated differencing can be done with *PEST* by selecting Option 8 from the Data Menu. *PEST* will only difference data that are stored in a file (since the undifferenced data is not recoverable from the differences only). If you wish to difference after making a Box-Cox transformation you will be required to file the transformed data before differencing. The program will instruct you to do this using Option 10 of the Data Menu.

> EXAMPLE: At this stage of the analysis you should have the original data set AIRPASS stored in *PEST* with the Data Menu displayed on the screen. Type 5↩ 0↩ to replace the stored observations by their natural logs. If you attempt to difference by typing 8↩ you will see the following instructions displayed on the screen:
>
> YOU MUST FILE THE TRANSFORMED SERIES BEFORE YOU WILL
> BE ALLOWED TO DIFFERENCE THE DATA SET. THIS WILL
> ALLOW YOU TO UN-DIFFERENCE LATER. PROCEED AS
> FOLLOWS :
>
>   PRESS ANY KEY
>
>   SELECT OPTION 10
>
>   SPECIFY A FILE NAME FOR THE TRANSFORMED SERIES
>
> Follow these instructions, typing AIRPASS.LOG↩ when asked for the file name. The transformed data will then be stored both on disk and in *PEST* under the name AIRPASS.LOG. The series AIRPASS.LOG can now be deseasonalized by differencing at lag 12. To do this type 8↩ 12↩ . Inspection of the graph of the deseasonalized series suggests a further differencing at lag 1 to eliminate the remaining trend. To do this type 8↩ 1↩ . Then

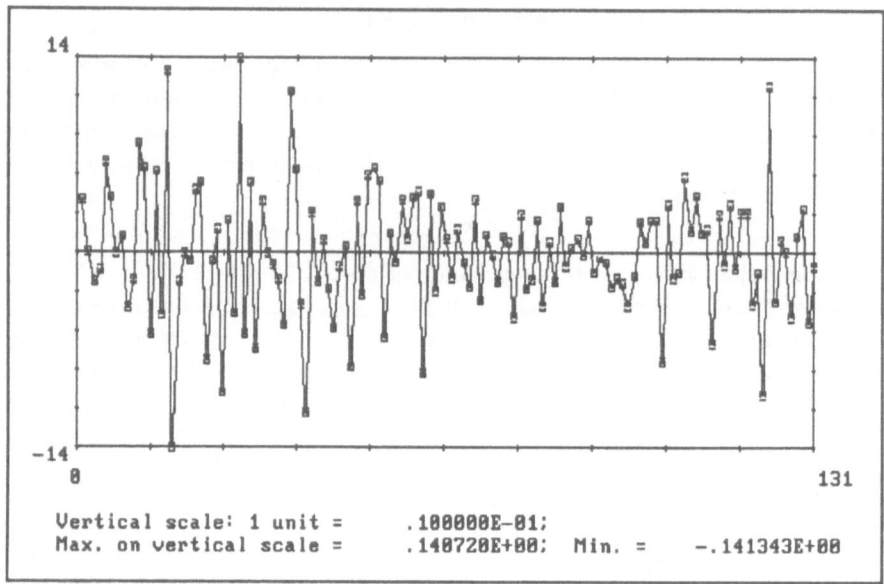

FIGURE 2.8. *The AIRPASS data after taking logs and differencing at lags 12 and 1*

type 2↩ ↩ and you should see the transformed and twice differenced series shown in Figure 2.8.

SUBTRACTING THE MEAN

The term *ARMA* model is used in this manual (and in *BD*) to mean a stationary zero mean process satisfying the defining difference equations in Section 2.6.1. In order to fit such a model to data, the sample-mean of the data should therefore be small. Once the apparent deviations from stationarity of the data have been removed, we therefore (in most cases) subtract the sample mean of the transformed data from each observation to generate a series to which we then fit a zero-mean stationary model. Effectively we are estimating the mean of the model by the sample mean, then fitting a (zero-mean) ARMA model to the *mean-corrected* transformed data. If we know a priori that the observations are from a process with zero mean then this process of mean correction is omitted. *PEST* keeps track of all the transformations (including mean correction) which are made. You can check these for yourself by going to Option 10 of the Main Menu.

When it comes time to predict the original series, *PEST* will invert all these transformations automatically.

> EXAMPLE: Subtract the mean of the transformed and twice differenced AIRPASS data by typing 9↵ . Type 11↵ to return to the Main Menu, then 10↵ to check the status of the data and model which currently reside in *PEST*. You will see in particular that no model has yet been entered.

2.3   Finding a Model for Your Data

After transforming the data (if necessary) as described in Section 2.2.4, we are now in a position to fit a zero-mean stationary time series model. *PEST* restricts attention to ARMA models (see Section 2.6.1). These constitute a very large class of zero-mean stationary time series. By appropriate choice of the parameters of an ARMA process $\{X_t\}$, we can arrange for the covariances $\text{Cov}(X_{t+h}, X_t)$ to be arbitrarily close, for all h, to the corresponding covariances $\gamma(h)$ of any stationary series with $\gamma(0) > 0$ and $\lim_{h\to\infty} \gamma(h) = 0$. But how do we find the most appropriate ARMA model for a given series? *PEST* uses a variety of tools to guide us in the search. These include the ACF (autocorrelation function), the PACF (partial autocorrelation function) and the AICC statistic (a bias-corrected form of Akaike's AIC statistic, see *BD Section 9.3*).

2.3.1   THE ACF AND PACF (*BD Sections 1.3, 3.3, 3.4, 8.2*)

The **autocorrelation function** (ACF) of the stationary time series $\{X_t\}$ is defined as

$$\rho(h) = \text{Corr}(X_{t+h}, X_t) \quad \text{for } h = 0, \pm 1, \pm 2, \dots$$

(Clearly $\rho(h) = \rho(-h)$ if X_t is real-valued, as we assume throughout.)

The ACF is a measure of dependence between observations as a function of their separation along the time axis. *PEST* estimates this function by computing the **sample autocorrelation function**, $\hat{\rho}(h)$ of the data $x_1, \dots, x_n$, i.e.

$$\hat{\rho}(h) = \hat{\gamma}(h)/\hat{\gamma}(0), \ 0 \le h < n,$$

where $\hat{\gamma}(\cdot)$ is the **sample autocovariance function**,

$$\hat{\gamma}(h) = n^{-1} \sum_{j=1}^{n-h} (x_{j+h} - \bar{x})(x_j - \bar{x}), \ 0 \le h < n.$$

Option 3 of the Data Menu can be used to compute and plot the sample ACF for values of the lag h from 1 up to 40. Values which decay rapidly as

h increases indicate short term dependency in the time series, while slowly decaying values indicate long term dependency. For ARMA fitting it is desirable to have a sample ACF which decays fairly rapidly (see *BD Chapter 9*). A sample ACF which is positive and very slowly decaying suggests that the data may have a trend. A sample ACF with very slowly damped periodicity suggests the presence of a periodic seasonal component. In either of these two cases you may need to transform your data before continuing (see Section 2.2.4).

Another useful diagnostic tool is the **sample partial autocorrelation function** or sample PACF.

The partial autocorrelation function (PACF) of the stationary time series $\{X_t\}$ is defined (at lag $h > 0$) as the correlation between the residuals of X_{t+h} and X_t after linear regression on $X_{t+1}, X_{t+2}, \ldots, X_{t+h-1}$. This is a measure of the dependence between X_{t+h} and X_t after removing the effect of the intervening variables $X_{t+1}, X_{t+2}, \ldots, X_{t+h-1}$. The sample PACF is estimated from the data $x_1, \ldots, x_n$ as described in *BD Section 3.4*.

The sample ACF and PACF are computed and plotted by choosing Option 3 of the Data Menu. *PEST* will prompt you to specify the maximum lag required. This is restricted by *PEST* to be less than or equal to 40. (As a rule of thumb, the estimates are reliable for lags up to about $\frac{1}{3}$ of the sample size. It is clear from the definition of the sample ACF, $\hat{\rho}(h)$, that it will be a very poor estimator of $\rho(h)$ for h close to the sample size n.)

Once you have specified the maximum lag, M, the sample ACF and PACF values will be plotted on the screen for lags h from 0 to M. The horizontal lines on the graph display the bounds $\pm 1.96/\sqrt{n}$ which are approximate 95% bounds for the autocorrelations of a white noise sequence. If the data is a (large) sample from an independent white noise sequence, approximately 95% of the sample autocorrelations should lie between these bounds. Large or frequent excursions from the bounds suggest that we need a model to explain the dependence and sometimes suggest the kind of model we need (see below). Press any key and the numerical values of the sample ACF and PACF will be printed below the graphs. Press any key again to return to the Data Menu.

The ACF and PACF may be filed for later use using Option 4.

The graphs of the sample ACF and PACF sometimes suggest an appropriate ARMA model for the data.

Suppose that the data are in fact observations of the MA(q) process,

$$X_t = Z_t + \theta_1 Z_{t-1} + \cdots + \theta_q Z_{t-q}.$$

The ACF of $\{X_t\}$ vanishes for lags greater than q and so the plotted sample ACF of the data should be negligible (apart from sampling fluctuations) for lags greater than q. As a rough guide, if the sample ACF falls between the plotted bounds $\pm 1.96/\sqrt{n}$ for lags $h > q$ then an MA(q) model is suggested.

Analogously, suppose that the data are observations of the AR(p) process

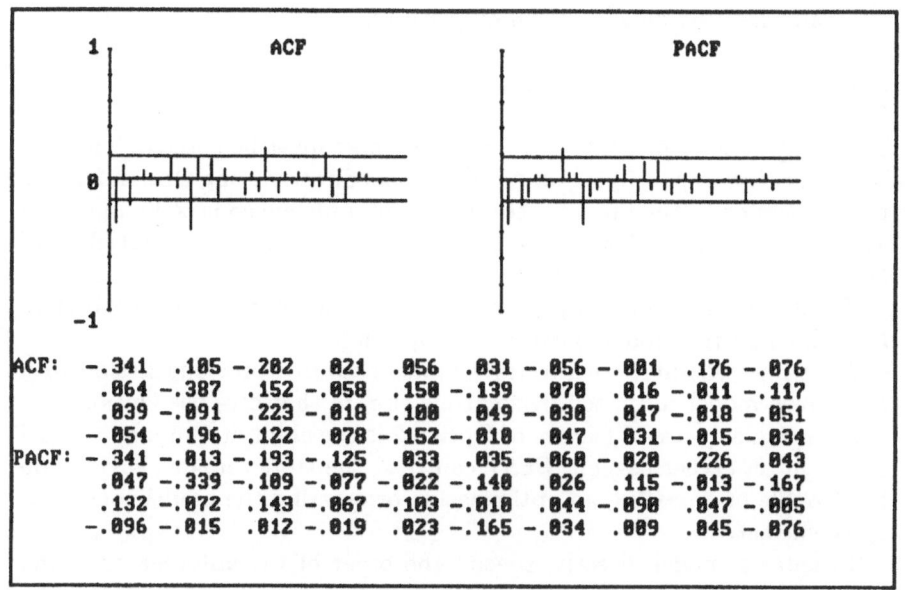

```
ACF:  -.341  .105 -.202  .021  .056  .031 -.056 -.001  .176 -.076
       .064 -.387  .152 -.058  .150 -.139  .070  .016 -.011 -.117
       .039 -.091  .223 -.018 -.100  .049 -.030  .047 -.018 -.051
      -.054  .196 -.122  .078 -.152 -.010  .047  .031 -.015 -.034
PACF: -.341 -.013 -.193 -.125  .033  .035  .060 -.020  .226  .043
       .047 -.339 -.109 -.077 -.022 -.148  .026  .115 -.013 -.167
       .132 -.072  .143 -.067 -.103 -.010  .044 -.090  .047 -.005
      -.096 -.015  .012 -.019  .023 -.165 -.034  .009  .045 -.076
```

FIGURE 2.9. *Sample ACF and PACF of the transformed AIRPASS data*

defined by
$$X_t = \phi_1 X_{t-1} + \cdots + \phi_p X_{t-p} + Z_t.$$
The PACF of $\{X_t\}$ vanishes for lags greater than p and so the plotted sample PACF of the data should be negligible (apart from sampling fluctuations) for lags greater than p. As a rough guide, if the sample PACF falls between the plotted bounds $\pm 1.96/\sqrt{n}$ for lags $h > p$ then an AR(p) model is suggested.

If neither the sample ACF nor PACF 'cuts off' as in the previous two paragraphs, a more refined model selection technique is required (see the discussion of the AICC statistic in Section 2.3.4 below). Even if the sample ACF or PACF does cut off at some lag, it is still advisable to explore models other than those suggested by the sample ACF and PACF values.

EXAMPLE: Figure 2.9 shows the ACF and PACF for the AIRPASS data after taking logarithms, differencing at lags 12 and 1 and subtracting the mean.

These graphs suggest we consider an MA model of order 12 (or perhaps 23) with a large number of zero coefficients, or alternatively an AR model of order 12.

2.3.2   ENTERING A MODEL

To do any serious analysis with *PEST*, a model must be entered. This can be done either by specifying an ARMA model directly using Option 2 of the Main Menu or (if the program contains a data file which is to be modelled as an ARMA process) by using the Preliminary Estimation procedure of Option 3.

If you have data and no particular ARMA model in mind, it is best to let *PEST* find the model. That is, use Option 3.

Sometimes you may wish to try a model used in a previous session with *PEST* or a model suggested by someone else. In that case use Option 2.

A particularly useful feature of Option 2 is the ability to import a model stored in an earlier session. *PEST* can read the stored model, saving you the trouble of repeating an optimization or entering the model coefficient by coefficient.

To enter a model directly, specify the order of the autoregressive and moving average polynomials as requested. You will then be required to enter the coefficients. Initially *PEST* will set the white noise variance to 1. To enter a model stored in a file, choose the autoregressive order to be −1.

After you have entered the model, you will see the Model Menu which gives you the opportunity to make any required changes.

If you wish to alter a specific coefficient in the model, enter the number of the coefficient. The autoregressive coefficients are numbered 1, 2, ..., p and the moving average coefficients are numbered $p+1, p+2, ..., p+q$. For example, to change the 2nd moving average coefficient in an ARMA(3,2) model, type 5↩ .

2.3.3   PRELIMINARY PARAMETER ESTIMATION (*BD Sections 8.1–8.5*)

Option 3 of the Main Menu provides you with a fast (but somewhat rough) modelling procedure. It is useful for exploring models to find which are the most promising for the data. More refined estimation should always be done using the slower but more efficient Option 8. A model fitted by Option 3 is useful as an initial approximation for starting the non-linear optimization carried out in Option 8. (For observations from a pure autoregressive series ($q = 0$), Option 3 gives estimates which are nearly as accurate for large samples as those from Option 8.)

The required AR and MA orders p and q must be entered first (see Section 2.6.1). If $q > 0$, *PEST* will then ask you to specify the value of m. This is a parameter required in the estimation algorithm (discussed in *BD Sections 8.3–8.4*). If you type 0↩ , *PEST* will make the choice for you.

Once the values of p, q and m have been entered, *PEST* will quickly estimate the parameters of the specified model and display a number of useful diagnostic statistics.

The estimated parameters are given with the ratio of each estimate to 1.96 times its standard error. The denominator (1.96 × standard error) is the critical value for the coefficient. Thus if the ratio is greater than one in absolute value, we may conclude (at level 0.05) that the corresponding coefficient is different from zero. On the other hand, a ratio less than one in absolute value suggests the possibility that the corresponding coefficient in the model may be zero.

After the estimated coefficients are displayed on the screen, press any key and *PEST* will then do one of two things depending on whether or not the fitted model is causal (see Section 2.6.1).

If the model is causal, *PEST* will give an estimate $\hat{\sigma}^2$ of the white noise variance, $\text{Var}(Z_t)$, and some further diagnostic statistics. These are $-2\ln L(\hat{\boldsymbol{\phi}}, \hat{\boldsymbol{\theta}}, \hat{\sigma}^2)$, where L denotes the Gaussian likelihood (see *BD equation (8.7.4)*), and the AICC statistic,

$$-2\ln L + 2(p+q+1)n/(n-p-q-2),$$

(see Section 2.3.4 below).

Our eventual aim is to find a model with as small an AICC value as possible. Smallness of the AICC value computed in Option 3 is indicative of a good model, but should be used only as a rough guide. Final decisions between models should be based on maximum likelihood estimation (Option 8), since for fixed p and q, the values of ϕ, θ and σ^2 which minimize the AICC statistic are the maximum likelihood estimates, not the preliminary estimates.

If the preliminary fitted model is non-causal, *PEST* will set all coefficients to .001 to generate a causal model with the specified values of p and q. Further investigation of this model must then be done with Option 8.

After completing the preliminary estimation, *PEST* will store the fitted model coefficients and white noise variance. The stored estimate of the white noise variance is the sum of squares of the residuals (or one-step prediction errors) divided by the number of observations.

At this point you can try a different model, file the current model or return to the Main Menu. When you return to the Main Menu, the most recently fitted preliminary model will be stored in *PEST*. You will now see a large number of options available on the Main Menu.

> EXAMPLE: Let us fit a preliminary MA(25) model to the logged, differenced and mean-corrected AIRPASS data currently stored in *PEST* under the name AIRPASS.LOG. Choose Option 3 from the Main Menu. Type 0↵ for the order of the autoregressive polynomial. Type 25↵ for the order of the moving average

```
   Order of autoregression to be fitted (p<27)? 0

   Order of moving average to be fitted (q<27)? 25

   Maximum lag, m, for autocovariances
     [p+q-1 < m < min(75, # of data values)]?
   Enter value of m (or 0 for automatic selection) : 0
MA COEFFICIENTS
   -.3567585       .0673203      -.1628929      -.0414917       .1267971
    .0264303       .0282778      -.0647944       .1326293      -.0761576
   -.0066282      -.4987470       .1700694      -.0317712       .1475751
   -.1468600       .0439758      -.0225709      -.0748717      -.0455962
   -.0204091      -.0085378       .2013022      -.0767226      -.0789432
RATIO OF COEFFICIENTS TO (1.96*STANDARD ERROR)
   -2.0833100       .3702632      -.0941187      -.2251242       .6874609
    .1423154       .1522179      -.3486668       .7124264      -.4060745
   -.0352567     -2.6528780       .8623497      -.1522100       .7067654
   -.6944412       .2076198      -.1064952      -.3532042      -.2147861
   -.0960386      -.0401670       .9474891      -.3563149      -.3659468
<Press any key to continue>
```

FIGURE 2.10. *The preliminary model estimate*

polynomial and type 0↩ for automatic selection of m, the number of autocovariances used in the estimation procedure.

The ratios, (estimated coefficient)/(1.96×standard error), indicate that the coefficients at lags 1 and 12 are non-zero, as we suspected from the ACF. The estimated coefficients at lags 3 and 23 also look substantial even though the corresponding ratios are less than 1 in absolute value.

The displayed values are shown in Figure 2.10. Press any key to see the values of $-2\ln L$ and the AICC.

To return to the Main Menu type 3↩ . The fitted MA(25) model is now stored in *PEST*.

2.3.4 THE AICC STATISTIC (*BD Sections 9.2, 9.3*)

One measure of the "goodness of fit" of a model is the Gaussian likelihood of the observations under the fitted model. (i.e. the joint probability density, evaluated at the observed values, of the random variables $X_1, \ldots, X_n$,

assuming that the fitted model is correct and has Gaussian white noise.)
At first glance, maximization of the Gaussian likelihood seems a plausible
criterion for deciding between rival candidates for "best" model to represent a given data set. For fixed p and q, maximization of the (Gaussian)
likelihood is indeed a good criterion and is the primary method used for
estimation in Option 8 of the Main Menu.

The problem with using the likelihood to choose between models of different orders is that for any given model, we can always find one with
equal or greater likelihood by increasing either p or q. For example, given
the maximum likelihood AR(10) model for a given data set, we can find an
AR(20) model for which the likelihood is at least as great. Any improve-
ment in the likelihood however is offset by the additional estimation errors
introduced. The AICC statistic allows for this by introducing a penalty
for increasing the number of model parameters. The AICC statistic for the
model with parameters p, q, ϕ, θ, and σ^2 is defined as

$$AICC(\phi, \theta, \sigma^2) = -2\ln L(\phi, \theta, \sigma^2) + 2(p + q + 1)n/(n - p - q - 2),$$

and a model chosen according to the AICC criterion minimizes this statistic. (The AICC value is a bias-corrected modification of the AIC statistic,
$-2\ln L + 2(p + q)$, see **BD Section 9.3.**)

Model selection statistics other than AICC are also available. A Bayesian
modification of the AIC statistic, known as the BIC statistic is also computed in Option 8. It is used in the same way as the AICC.

An exhaustive search for a model with minimum AICC or BIC value
can be very slow. For this reason the sample ACF and PACF and the
preliminary estimation techniques in Option 3 of the Main Menu are useful
in narrowing down the range of models to be considered more carefully in
Option 8.

2.3.5   CHANGING YOUR MODEL

The model currently stored by the program and the status of the data file
can be checked at any time using Option 10 of the Main Menu. Any parameter can be changed with this option, including the white noise variance,
and the model can be filed for use at some other time.

> EXAMPLE: We shall now set some of the coefficients in the
> current model to zero. To do this choose Option 10 of the Main
> Menu. The resulting screen display is shown in Figure 2.11.
>
> The preliminary estimation in Section 2.3.3 suggested that the
> most significant coefficients in the fitted MA(25) model were
> those at lags 1, 3, 12 and 23. Let us therefore try setting all the
> other coefficients to zero. To change the lag-2 coefficient, enter
> its number followed by the new value, 0, i.e. type 2↩ 0↩ .

```
   The datafile is airpass.log              ;  Total data points=   131

   Box-Cox transform of prior data set, lambda =    .00

   Difference     lag
        1         12
        2          1


   The subtracted mean is           .0003

THE ARMA( 0,25) MODEL IS  X(t) = Z(t)
 +(  -.357)*Z(t- 1)  +(   .067)*Z(t- 2)  +(  -.163)*Z(t- 3)  +(  -.041)*Z(t- 4)
 +(   .127)*Z(t- 5)  +(   .026)*Z(t- 6)  +(   .028)*Z(t- 7)  +(  -.065)*Z(t- 8)
 +(   .133)*Z(t- 9)  +(  -.076)*Z(t-10)  +(  -.007)*Z(t-11)  +(  -.499)*Z(t-12)
 +(   .178)*Z(t-13)  +(  -.032)*Z(t-14)  +(   .148)*Z(t-15)  +(  -.146)*Z(t-16)
 +(   .044)*Z(t-17)  +(  -.023)*Z(t-18)  +(  -.075)*Z(t-19)  +(  -.046)*Z(t-20)
 +(  -.020)*Z(t-21)  +(  -.009)*Z(t-22)  +(   .201)*Z(t-23)  +(  -.077)*Z(t-24)
 +(  -.079)*Z(t-25)            *MODEL NOT INVERTIBLE*


White noise variance =    .115169E-02
   <Press any key to continue>
```

FIGURE 2.11. *The PEST screen after choosing Option 10*

Repeat for each coefficient to be changed. The screen should then look like Figure 2.12. Type 0↵ to return to the Main Menu.

2.3.6 PARAMETER ESTIMATION; THE GAUSSIAN LIKELIHOOD (*BD Section 8.7*)

Once you have specified values of p and q and possibly set some coefficients to zero, you can use the full power of *PEST* to estimate parameters. For efficient parameter estimation you must use Option 8.

From the Main Menu type 8↵ to obtain the Estimation Menu displayed in Figure 2.13.

Much of the information displayed in this menu concerns the optimization settings. For most purposes you will need to use the default settings only. (With the default settings, any coefficients which are set to zero will be treated as fixed values and not as parameters. If you wish to include a particular coefficient in the parameters to be optimized you must therefore not set its initial value equal to zero.)

```
THE ARMA( 0,25) MODEL IS  X(t) = Z(t)
 +(  -.357)*Z(t- 1)  +(  -.163)*Z(t- 3)  +(  -.499)*Z(t-12)  +(   .201)*Z(t-23)


White noise variance =    .115172E-02
   <Press any key to continue>


Parameters to be changed ?
      -3 to change the white noise variance
      -2 to file the model
      -1 to input different model
       0 for none
       n (>0) to change the nth coeff, e.g. enter p+2 for theta(2)
Choose a number:
```

FIGURE 2.12. *The PEST screen after setting coefficients to zero*

To find the maximum likelihood estimates of your parameters choose Option 1 (Optimize with Current Settings). *PEST* will then try to find the parameters which maximize the likelihood of your model with respect to all the non-zero coefficients in the model currently stored by *PEST*.

If you wish to compute the Gaussian likelihood (and one-step predictors) without doing any optimization, choose Option 0 of the Estimation Menu by typing 0↩ .

> EXAMPLE: Find the maximum likelihood estimates of the parameters in the current model for the transformed, differenced and mean-corrected airline passenger data stored in *PEST* as AIRPASS.LOG. Starting from the Main Menu type 8↩ and you will see the Estimation Menu. Choose the default option by typing 1↩ . After a short delay the iterations will cease and you will see the message
>
> STOPPING VALUE 2 : WITHIN ACCURACY LEVEL
>
> This indicates that the minimum of $-2\ln L$ has been located with the specified accuracy. The fitted model is displayed in

```
            P A R A M E T E R   E S T I M A T I O N
THE ARMA( 0,25) MODEL IS  X(t) = Z(t)
 +(  -.357)*Z(t- 1)  +(  -.163)*Z(t- 3)  +(  -.499)*Z(t-12)  +(   .201)*Z(t-23)

   Method : MAXIMUM LIKELIHOOD
   Max. No. of Iterations:      7,  Accuracy Parameter:   .00205254
   OPTIMIZING  4 PARAMETERS          THE NON-ZERO COEFFICIENTS
ESTIMATION MENU :
  -1: Help
   0: Likelihood of Model (no optimization)
   a: (0<a<1) Reset Accuracy Parameter to a and Optimize
   1: Optimize with Current Settings
   2: Change Accuracy Parameter
   3: Change Maximum No. of Iterations
   4: Change Optimized Coefficients (e.g. for Multiplicative Model)
   5: Change Convergence Criterion for th(n,j):currently  .00005000
   6: Change Method to LEAST SQUARES
   7: Return to Main Menu
   n: (where n >7) Optimize with at Most n Iterations
CHOOSE A NUMBER : 1
 ITERATION
```

FIGURE 2.13. *The Estimation Menu*

Figure 2.14. If you see the message

STOPPING VALUE 4 : ITERATION LIMIT EXCEEDED

then the minimum of $-2 \ln L$ could not be located with the number of iterations (7) allowed. You can continue the search (starting from the point at which the iterations were interrupted) by typing 4↩ to return to the Estimation Menu, then typing 1↩ as before.

CHANGING THE OPTIMIZATION SETTINGS

There are several options in the Estimation Menu which enable you to alter the way in which the optimization is carried out. In particular, it is possible to reset the accuracy parameter a (Option 2), the maximum number of iterations m (Option 3), the convergence criterion c (Option 5) and the method of optimization (Option 6). By far the most frequently used option is 1, however it is a good idea to conclude the estimation with a further optimization using a smaller accuracy parameter.

The following options on the Estimation Menu can be used to alter the

```
THE ARMA( 0,25) MODEL IS  X(t) = Z(t)
 +(  -.355)*Z(t- 1)  +(  -.201)*Z(t- 3)  +(  -.524)*Z(t-12)  +(   .242)*Z(t-23)

   MOVING AVERAGE PARAMETERS :
     THETA( 1)=              -.35502100        THETA( 3)=            -.20127600
     THETA(12)=              -.52359250        THETA(23)=             .24187590

   WHITE NOISE VARIANCE =            .125004E-02
   BIC STATISTIC         =          -487.633300
   -2 ln(LIKELIHOOD)     =          -496.517500
   AICC STATISTIC        =          -486.037500

    # FUNCTION CALLS =       46 ;# ITERATIONS =    5;ACCURACY PARAM.=  .002053
    STOPPING VALUE  2 : WITHIN ACCURACY LEVEL

    CONVERGENCE OCCURRED AT STEP  100 OF  131; VALUE OF R(N) =       1.0023670
RESULTS MENU :
     1. Store Model       2. File and Analyze Residuals
     3. Prediction
     4. Parameter Optimization with this Model
     5. Redo Screen with Standard Errors
     6. Show Covariance Matrix of Optimized Parameters
     8. Return to Main Menu
CHOOSE A NUMBER :
```

FIGURE 2.14. *The maximum likelihood estimates for the transformed AIRPASS data*

optimization settings.

Option a

Entering a number between 0 and 1 sets the accuracy parameter a to the number entered and optimizes the coefficients accordingly. Reducing a gives more accurate optimization.

Option 2

Changes the accuracy parameter but does not automatically begin optimizing.

Option 3

Changes the maximum number of iterations m required before terminating the search. Reducing m terminates the search after fewer iterations.

Option 4

Unless you specify otherwise, *PEST* will optimize only the non-zero coefficients in the current model. Sometimes you may not want this. Option 4 enables you to specify the coefficients to be optimized. Coefficients can be set to non-zero constant values and coefficients which are currently zero can be treated as parameters and included in the optimization. It is also possible to specify optimization subject to multiplicative relationships between the parameters (see below under Multiplicative Models).

Option 5

Changes the convergence criterion c. Setting $c = 0$ gives the exact likelihood but setting c to be small (say 0.0005) will usually give an almost identical value with far less computation.

Option 6

Toggles the method of optimization between Maximum Likelihood (the default option) and Least Squares.

Option n

Entering a number greater than 7 sets the maximum number of iterations m to the number entered and causes *PEST* to begin optimizing.

STOPPING NUMBER

Each time the optimizing iterations cease, the resulting model will be displayed on the screen with a "stopping number" indicating the conditions under which the search was terminated. The stopping numbers have the following meanings:

1. The relative gradient of the surface is close to zero.

2. Successive iterations did not change any of the optimized parameters by more than the required accuracy parameter a.

3. The last step failed to locate a better point. Either the value is an approximate local minimum or the model is too non-linear at this point — perhaps because of a root near the unit circle. Try different initial values.

4. The iteration limit (m) was reached. (Continue optimization by entering 4↩ from the results screen then entering 1↩ again.)

5. The step-size of the search has grown too large. Try different initial values.

   EXAMPLE: In the optimization just completed, stopping number 2 appeared after 6 iterations.

MULTIPLICATIVE MODELS (*BD Section 9.6*)

Option 4 of the Estimation Menu allows the imposition of more complicated constraints on the parameters. Multiplicative models are handled by specifying multiplicative relationships between the ARMA coefficients. For example the multiplicative ARMA model,

$$(1 - aB)X_t = (1 + cB)(1 + dB^{12})Z_t,$$

is fitted as follows. After entering the data (and transforming if necessary), use Option 2 of the Main Menu to enter the ARMA(1,13) model with all coefficients zero except those with indices 1, 2, 13 and 14. These may be set initially to .001 (or some better non-zero initial values obtained for example from Option 3). We then choose Option 8 and enter the following sequence of numbers:

4↩ (Change Optimized Coefficients),

5↩ (Define Multiplicative Relationships),

1↩ (Multiplicative Relationship),

2↩ 13↩ 14↩ (the 14th coefficient is constrained to be the product of the 2nd and 13th),

1↩ (Return to OPTIMIZATION), and finally

1↩ (Optimize with Current Settings).

HINTS FOR ADVANCED USERS

- You cannot specify more than 40 coefficients to be optimized.

- If the optimization search takes the MA coefficients outside the invertible region you can convert the model to an equivalent (from a second-order point of view) invertible model. See Section 2.3.7 for further information about switching to invertible models. *PEST* (unlike some programs) has no difficulty in computing Gaussian likelihoods and best linear predictors for non-invertible models.

- For complicated or high-order models be sure to try a variety of initial values to check that you are not finding a local rather than a global minimum of $-2 \ln L$.

- You cannot keep constant the 3rd coefficient in a multiplicative relationship (i.e. the product of the first 2).

2.3.7 OPTIMIZATION RESULTS

After running the optimization algorithm, *PEST* displays the model parameters (coefficients and white noise variance), the values of $-2 \ln L$, AICC, and BIC, information regarding the computations, and the Results Menu.

> EXAMPLE: Figure 2.14 shows the *PEST* screen after completing optimization for the logged, differenced and mean-corrected data AIRPASS.LOG, using an MA(23) model with non-zero coefficients at lags 1, 3, 12 and 23.

The next stage of the analysis is to consider a variety of competing models and to select the most suitable. The following table shows the AICC statistics for a variety of subset moving average models of order less than 24.

Lags						AICC
1	3		12		23	-486.04
1	3		12	13	23	-485.78
1	3	5	12		23	-489.95
1	3		12	13		-482.62
1			12			-475.91

The best of these models from the point of view of AICC value is the one with non-zero coefficients at lags 1, 3, 5, 12 and 23. To substitute this model for the one currently stored in *PEST* (starting from the Results Menu), type

$$4 \hookleftarrow 4 \hookleftarrow 9 \hookleftarrow 1 \hookleftarrow 5 \hookleftarrow 1 \hookleftarrow 1 \hookleftarrow$$

and optimization will begin as before. You should obtain the non-invertible model (*BD Example 9.2.2*),

$$X_t = Z_t - .439Z_{t-1} - .302Z_{t-3} + .242Z_{t-5}$$
$$-.656Z_{t-12} + .348Z_{t-23}, \quad \{Z_t\} \sim \text{WN}(0, .00103)$$

For future reference, store the model under the filename AIR-PASS.MOD using Option 1 of the Results Menu.

We have seen in our example how, when optimizing iterations cease, the stopping code indicates whether or not more iterations are required. The display of results which includes this information also contains a menu which allows us to investigate properties of the fitted model, including its goodness of fit. The options available in the Results Menu are described below.

STORE MODEL

It is wise to file the fitted model, particularly if it was the result of a time-consuming optimization as in the previous example.

FILE AND ANALYZE RESIDUALS

The differences (suitably rescaled, see *BD Section 9.4*) between the observations and the corresponding one-step predictors are the residuals from the model. If the fitted model were the true model, they would constitute a white noise sequence. This allows us to check, by studying the residuals, whether or not the model is a good fit to the data. Further details of this option are given in Section 2.4.

```
THE ARMA( 0,25) MODEL IS  X(t) = Z(t)
 +(  -.439)*Z(t- 1)  +(  -.302)*Z(t- 3)  +(   .241)*Z(t- 5)  +(  -.656)*Z(t-12)
 +(   .348)*Z(t-23)

    MOVING AVERAGE PARAMETERS :        STANDARD ERROR
    THETA( 1)=              -.4386914       .1026168
    THETA( 3)=              -.3021560       .0793478
    THETA( 5)=               .2413256       .0865697
    THETA(12)=              -.6559475       .0817836
    THETA(23)=               .3482866       .0922978
RESULTS MENU :
    1. Store Model       2. File and Analyze Residuals
    3. Prediction
    4. Parameter Optimization with this Model
    5. Return to Original Results Page
    6. Show Covariance Matrix of Optimized Parameters
    7. NONINVERTIBLE MODEL. Switch to Invertible Model
    8. Return to Main Menu
CHOOSE A NUMBER :
```

FIGURE 2.15. *The standard errors of the coefficient estimators*

PREDICTION

It is possible to compute best linear h-step predictors for both the transformed and original series using the fitted model and Option 3 of the Results Menu. This option is also available directly from the Main Menu and is discussed later in Section 2.6.

STANDARD ERRORS

The standard errors (estimated standard deviations) of the coefficient estimators can be printed on the screen by choosing Option 5 of the Results Menu. These are evaluated by numerical determination of the Hessian matrix of $-2\ln L$ (*BD Section 9.2*).

> EXAMPLE: Type 5↩ to obtain the standard errors for the coefficient estimators in the model, AIRPASS.MOD, just fitted to the differenced and mean-corrected AIRPASS.LOG data (see Figure 2.15).

```
THE ARMA( 0,25) MODEL IS  X(t) = Z(t)
  +(   -.439)*Z(t- 1)  +(   -.302)*Z(t- 3)  +(    .241)*Z(t- 5)  +(  -.656)*Z(t-12)
  +(    .348)*Z(t-23)


    COVARIANCE MATRIX OF THE  5
    OPTIMIZED PARAMETERS
 TH( 1)  TH( 3)  TH( 5)  TH(12)  TH(23)
   .011   -.004   -.003   -.001   -.001
  -.004    .006   -.001    .002   -.003
  -.003   -.001    .007   -.004    .000
  -.001    .002   -.004    .007   -.003
  -.001   -.003    .000   -.003    .009
RESULTS MENU :
    1. Store Model        2. File and Analyze Residuals
    3. Prediction
    4. Parameter Optimization with this Model
    5. Redo Screen with Standard Errors
    6. Return to Original Results Page
    7. NONINVERTIBLE MODEL. Switch to Invertible Model
    8. Return to Main Menu
CHOOSE A NUMBER :
```

FIGURE 2.16. *The covariance matrix of the coefficient estimators*

COVARIANCE MATRIX

The estimated covariance matrix of the coefficient estimators can be printed
on the screen by choosing Option 6 of the Results Menu. (The diagonal
elements are the squares of the standard errors found in Option 5).

> EXAMPLE: Type 6↩ to obtain the estimated covariances for
> the coefficient estimators in the model for the differenced and
> mean-corrected AIRPASS.LOG data (see Figure 2.16).

NON-INVERTIBLE MODELS

If *PEST* fits a non-invertible (Section 2.6.1) $ARMA(p, q)$ model to your
data set, you can convert to an equivalent invertible $ARMA(p, q)$ model
using Option 7 of the Results Menu. Equivalent here means from a second-
order point of view. Note however that a non-invertible subset $ARMA(p, q)$
model will generally convert to an invertible $ARMA(p, q)$ model with all
q moving average coefficients non-zero. Once the model is converted to an
invertible model (if it is important to do so), Options 5 and 6 (for computing

standard errors) disappear from the Results Menu. At this stage, one should reoptimize (type **4↩ 1↩**) with the invertible model to get the standard errors of the estimated parameters.

2.4 Testing Your Model (*BD Section 9.4*)

Once we have a model, it is important to check whether it is any good or not. Typically this is judged by comparing observations with corresponding predicted values obtained from the fitted model. If the fitted model is appropriate then the prediction errors should behave in a manner which is consistent with the model.

We define the **residuals** to be the rescaled one-step prediction errors,

$$\hat{W}_t = (X_t - \hat{X}_t)/\sqrt{r_{t-1}},$$

where $\hat{X}_t$ is the best linear mean-square predictor of X_t based on the observations up to time $t-1$, $r_{t-1} = E(X_t - \hat{X}_t)^2/\sigma^2$ and σ^2 is the white noise variance of the fitted model.

If the data were truly generated by the fitted ARMA(p, q) model with white noise sequence $\{Z_t\}$, then for large samples the properties of $\{\hat{W}_t\}$ should reflect those of $\{Z_t\}$ (see *BD Section 9.4*). To check the appropriateness of the model we can therefore examine the residual series $\{\hat{W}_t\}$, and check that it resembles a realization of a white noise sequence.

PEST provides a number of tests for doing this in the Residuals Menu which is obtained by selecting Option 2 (File and Analyze Residuals) of the Results Menu.

To examine the residuals from a *specified* model without doing any optimization, enter the data and model, then use Option 0 of the Estimation Menu followed by Option 2 of the Results Menu.

> EXAMPLE: Choose Option 2 (File and Analyze Residuals) of the Results Menu and you will see the screen display shown in Figure 2.17.

2.4.1 PLOTTING THE RESIDUALS

The residuals $\hat{W}_t$, $t = 1, \ldots, n$ were defined in Section 2.4. The rescaled residuals are defined as

$$\hat{W}_t^{(r)} = \sqrt{n}\hat{W}_t/(\sum_{j=1}^{n} \hat{W}_t^2).$$

If the fitted model is appropriate, the histogram of the rescaled residuals should have mean close to zero. If the fitted model is appropriate and the

```
RESIDUALS MENU:
   1. File residuals
   2. Plot rescaled residuals
   3. Plot ACF/PACF of residuals
   4. File ACF/PACF of residuals
   5. Tests of randomness of residuals
   6. Return to result page
CHOOSE A NUMBER :
```

FIGURE 2.17. *The Residuals Menu*

data is Gaussian, this will be reflected in the shape of the histogram, which should then resemble a normal density with mean zero and variance one.

Press any key after inspecting the histogram and you will see a graph of $\hat{W}_t^{(r)}$ vs t. If the fitted model is appropriate this should resemble a realization of a white noise sequence. Look for trends, cycles and non-constant variance, any of which suggest that the fitted model is inappropriate. If substantially more than 5% of the rescaled residuals lie outside the bounds ± 1.96 or if there are rescaled residuals far outside these bounds, then the fitted model should not be regarded as Gaussian.

> EXAMPLE: Type 1↩ to choose Option 1 of the Residuals Menu. You will then see the histogram of the rescaled residuals as shown in Figure 2.18. The mean is close to zero and the shape suggests that the assumption of Gaussian white noise is not unreasonable in our proposed model for the transformed airline passenger data.

> Press any key to see the graph shown in Figure 2.19. A few of the rescaled residuals are greater in magnitude than 1.96 (as is

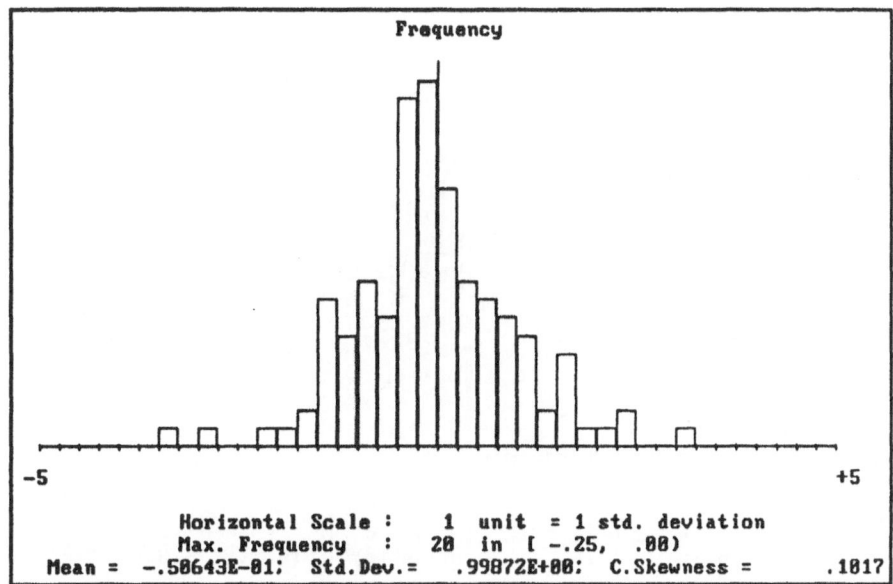

FIGURE 2.18. *Histogram of the rescaled residuals from AIRPASS.MOD*

to be expected), but there are no obvious indications here that the model is inappropriate. Press any key then type 0↩ to return to the Residuals Menu.

2.4.2 ACF/PACF OF THE RESIDUALS (*BD Section 9.4*)

If we were to assume that our fitted model is the true process generating the data, then the observed residuals would be realized values of a white noise sequence. We can check the hypothesis that $\{W_t\}$ is an independent white noise sequence by examining the sample autocorrelations of the observed residuals which should resemble observations of independent normal random variables with mean 0 and variance $1/n$ (see *BD Example 7.2.1*).

In particular the sample ACF of the observed residuals should lie within the bounds $\pm 1.96/\sqrt{n}$ roughly 95% of the time. These bounds are displayed on the graphs of the ACF and PACF. If substantially more than 5% of the correlations are outside these limits, or if there are a few very large values, then we should look for a better-fitting model. (More precise bounds, due to Box and Pierce, can be found in *BD Section 9.4*.)

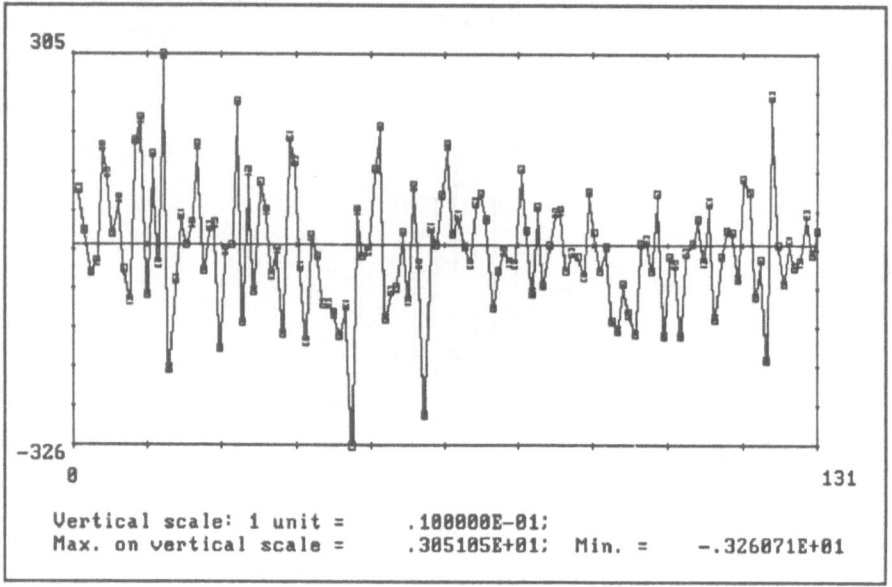

FIGURE 2.19. *Time plot of the rescaled residuals from AIRPASS.MOD*

EXAMPLE: Type 3↩ to choose Option 3 of the Residuals Menu. The sample ACF and PACF of the residuals will then appear as shown in Figure 2.20.

No correlations are outside the bounds in this case. They appear to be compatible with the hypothesis that the residuals are in fact observations of a white noise sequence. Type ↩ to return to the Residuals Menu.

2.4.3 TESTING FOR RANDOMNESS OF THE RESIDUALS (*BD Section 9.4*)

Option 5 of the Residuals Menu provides four tests of the hypothesis that the residuals are observations from an independent and identically distributed (iid) sequence.

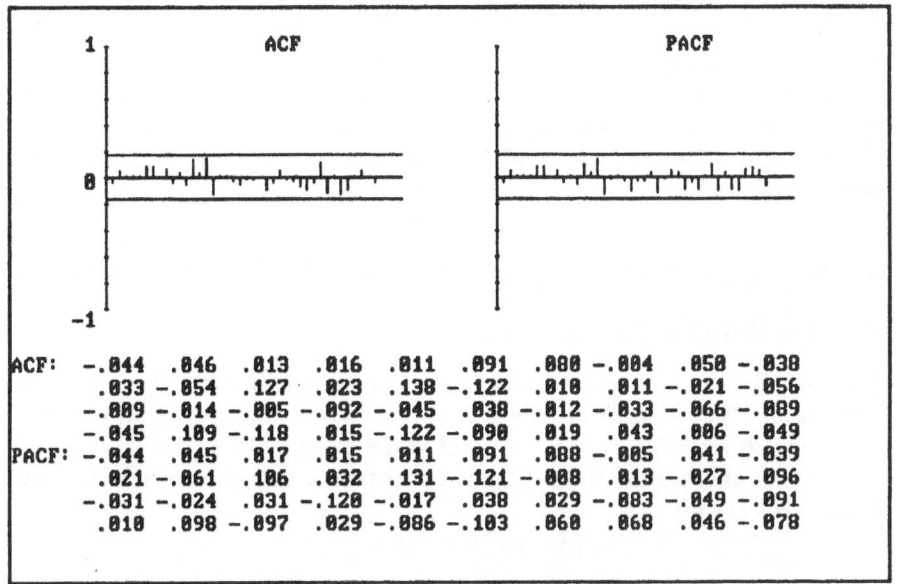

ACF: -.044 .046 .013 .016 .011 .091 .000 -.004 .050 -.038
       .033 -.054 .127 .023 .138 -.122 .010 .011 -.021 -.056
      -.009 -.014 -.005 -.092 -.045 .038 -.012 -.033 -.066 -.089
      -.045 .109 -.118 .015 -.122 -.090 .019 .043 .006 -.049
PACF: -.044 .045 .017 .015 .011 .091 .008 -.005 .041 -.039
       .021 -.061 .106 .032 .131 -.121 -.008 .013 -.027 -.096
      -.031 -.024 .031 -.120 -.017 .038 .029 -.003 -.049 -.091
       .010 .098 -.097 .029 -.086 -.103 .068 .068 .046 -.078

FIGURE 2.20. *ACF/PACF of the residuals from AIRPASS.MOD*

PORTMANTEAU TEST

This test, due to Box and Pierce, pools the sample autocorrelations of the residuals instead of looking at them individually. The statistic used is

$$Q = n \sum_{k=1}^{h} \hat{\rho}_W^2(k),$$

where $\hat{\rho}_W^2(k)$ is the sample autocorrelation of the residuals at lag k, and h is to be specified. As a rule of thumb, h should be of the order of $\sqrt{n}$, where n is the sample size ($h=20$ is a commonly used value).

If the data had in fact been generated by the fitted ARMA(p, q) model, then for large n, Q would have an approximate χ^2 distribution with $h - p - q$ degrees of freedom. The test rejects the proposed model at level α if the observed value of Q is larger than the $(1-\alpha)$ quantile of the χ^2_{h-p-q} distribution.

This test frequently fails to reject poorly fitting models. Care should be taken not to accept a model on the basis of the portmanteau test alone.

A TEST BASED ON TURNING POINTS

The statistic, T, used in this test is the number of turning points in the sequence of residuals. It can be shown that for an iid sequence, T is asymptotically normal with mean $\mu_T = 2(n-2)/3$ and variance $\sigma_T^2 = (16n-29)/90$.

The hypothesis that the residuals constitute a sequence of iid observations is rejected if

$$|T - \mu_T|/\sigma_T > \Phi_{1-\alpha/2},$$

where $\Phi_{1-\alpha/2}$ is the $(1-\alpha/2)$ quantile of the standard normal distribution.

THE DIFFERENCE-SIGN TEST

Let S be the number of times the differenced residual series $\hat{W}_t - \hat{W}_{t-1}$ is positive. If $\{\hat{W}_t\}$ is an iid sequence it can be shown that S is asymptotically normal with mean $\mu_S = \frac{1}{2}(n-1)$ and variance $\sigma_S^2 = (n+1)/12$.

The hypothesis that the residuals constitute a sequence of iid observations is rejected if

$$|S - \mu_S|/\sigma_S > \Phi_{1-\alpha/2}.$$

This test must be used with caution. If the residuals have a strong cyclic component they will be likely to pass the difference-sign test since roughly half of the differences will be positive.

THE RANK TEST

This test is particularly useful for detecting a linear trend in the residuals. Let P be the number of pairs (i, j) such that $\hat{W}_j > \hat{W}_i$, and $j > i$, $i = 1, \ldots, n-1$. If the residuals are iid, the mean of P is $\mu_P = \frac{1}{4}n(n-1)$, the variance of P is $\sigma_P^2 = n(n-1)(2n+5)/8$ and P is asymptotically normal.

The hypothesis that the residuals constitute a sequence of iid observations is rejected if

$$|P - \mu_P|/\sigma_P > \Phi_{1-\alpha/2}.$$

EXAMPLE: Type 5↵ to select Option 5 from the Residuals Menu. You will see the results shown in Figure 2.21. Every test is easily passed by our fitted model with $\alpha < .05$. Type ↵ to return to the Residuals Menu. For later use, save the residuals under the filename AIRPASS.RES using Option 1.

2.5 Prediction (BD Chapter 5, Section 9.5)

One of the main purposes of time series modelling is the prediction of future observations. Once you have found a suitable model for your data, you can predict future values using either Option 9 of the Main Menu or (equivalently) Option 3 of the Results Menu.

```
    RANDOMNESS TEST STATISTICS (see section 9.4)
    --------------------------------------------

PORT'TEAU with h=  20:     10.6049 chi sq df=  15

TURNING POINTS   =       87. ANORMAL(        86.00           4.79**2)

DIFFERENCE-SIGN =        65. ANORMAL(        65.00           3.32**2)

RANK TEST        =     3928. ANORMAL(      4257.50         753.91**2)


    <Press any key to continue>
```

FIGURE 2.21. *Tests of randomness for the residuals from AIRPASS.MOD*

2.5.1 FORECAST CRITERIA

Given observations $X_1, \ldots, X_n$ of a series which we assume to be appropriately modelled as an ARMA(p,q) process, *PEST* predicts future values of the series X_{n+h} from the data and the model by computing the linear combination $P_n(X_{n+h})$ of $X_1, \ldots, X_n$ which minimizes the mean squared error $E(X_{n+h} - P_n(X_{n+h}))^2$.

2.5.2 FORECAST RESULTS

Assuming that you have data stored in *PEST* which has been adequately fitted by an ARMA(p,q) model, also stored in *PEST*, choose Option 9 from the Main Menu, after which you will be asked for the number of future values you wish to predict.

After the model is displayed on the screen you will be asked if you wish to change the white noise variance. (This will not affect the predictors but only their mean squared errors.) The predicted values of the fitted ARMA process will then be displayed in the column labelled XHAT. In the column

labelled SQRT(MSE) you will see the square roots of the estimated mean
squared errors of the corresponding predictors. These are calculated under
the assumption that the observations are truly generated by the current
model. They measure the uncertainty of the corresponding forecasts. A
smaller value indicates a more reliable forecast. As is to be expected, the
mean squared error of $P_n(X_{n+h})$ increases with the lead time h of the
forecast.

Approximate 95% prediction bounds (*BD Section 5.4*) can be obtained
from each predicted value by adding and subtracting $1.96\sqrt{MSE}$. These
are exact under the assumptions that the model is Gaussian and faithfully
represents the data. They should not be interpreted as 95% bounds if the
histogram of the residuals is decidedly non-Gaussian in appearance.

If your data was mean-corrected, the third column of the *PEST* output
will show the predicted values in Column 1 plus the previously subtracted
sample mean. If there has been no mean-correction, the third column will
be the same as the first.

2.5.3   INVERTING TRANSFORMATIONS

The predictors and mean squared errors calculated so far do not pertain
to your *original* time series unless you have made no data transformations
other than mean-correction (in which case the relevant predictors are those
in the third column). What we have found are predictors of your *transformed* series. To predict the original series, you will need to invert all
the data transformations which you have made in order to fit a zero-mean
stationary model. *PEST* will do this for you. In fact one transformation,
mean-correction, has already been inverted to generate the predicted values
which were displayed in Column 3.

If you used differencing transformations, you will see the Prediction Menu
displayed following the printing of the ARMA predictors. Type 2↩ to
select the option Undo differencing. The predictors of the undifferenced data
(still Box-Coxed if you made such a transformation) will then be printed
on the screen together with the square roots of their mean squared errors.
Type ↩ and the undifferenced data will be plotted. Type ↩ again and the
predicted values will be added to the graph of the data. Type ↩ 0↩ and
you will be asked if you wish to invert the Box-Cox transformation (if you
made one). If so type y and the original data will be plotted on the screen.
Type ↩ again and the predictors of the original series will be added to the
graph. Type ↩ 0↩ and you will be asked if you wish to file the predicted
values, then returned to the Prediction Menu.

If you used classical decomposition rather than differencing, *PEST* will
automatically add back the trend and/or seasonal component immediately
after listing the ARMA predictors. The resulting data values and corresponding predictors will then be plotted, after which you will again be
given the opportunity to invert the Box-Cox transformation (if any) as in

```
+(   .348)=Z(t-23)

  MOVING AVERAGE PARAMETERS :
   THETA( 1)=            -.43869140      THETA( 3)=           -.30215600
   THETA( 5)=             .24132560      THETA(12)=           -.65594750
   THETA(23)=             .34828660

  WHITE NOISE VARIANCE =          .102770E-02
The WNV has been set to (RESIDUAL SS)/n =   .102770E-02

Do you wish to change it (y/n)?n

     #        XHAT           SQRT(MSE)         XHAT+MEAN (=   .29088E-03)
    132     .999148E-02     .342196E-01       .102824E-01
    133     .895318E-02     .364256E-01       .924406E-02
    134     .391665E-01     .364258E-01       .394573E-01
    135    -.481165E-01     .375163E-01      -.398256E-01
    136    -.124040E-01     .375920E-01      -.121139E-01
    137     .875554E-02     .379730E-01       .904642E-02
    138     .638590E-03     .379831E-01       .929476E-03
    139     .914851E-02     .380493E-01       .943939E-02
    140    -.168015E-02     .380775E-01      -.138928E-02
    141     .212363E-02     .380819E-01       .241450E-02
    142     .141257E-01     .380886E-01       .144166E-01
<Press any key to continue>
```

FIGURE 2.22. *Prediction of the transformed series, AIRPASS.LOG*

the previous paragraph.

   EXAMPLE: We left our logged, differenced and mean-corrected
   airline passenger data stored in *PEST* as AIRPASS.LOG along
   with the fitted MA(23) model, AIRPASS.MOD. To predict the
   next 24 values of the original series AIRPASS, first return to
   the Main Menu and select Option 9 (Prediction). Type 24↵
   to specify that 24 predicted values are required after the last
   observation. After a brief delay, you will be asked if you wish
   to change the white noise variance. Type n and the predictors
   will be displayed as in Figure 2.22. (Only the first 11 predicted
   points are shown.)

   To obtain forecasts of the undifferenced series, choose Option
   2 (Undo differencing) from the Prediction Menu and follow the
   program prompts to obtain the graph shown in Figure 2.23.
   Here the hollow squares represent the observations and the solid
   squares represent the forecast values. Notice how the model has
   captured the regular cyclic behaviour in the data.

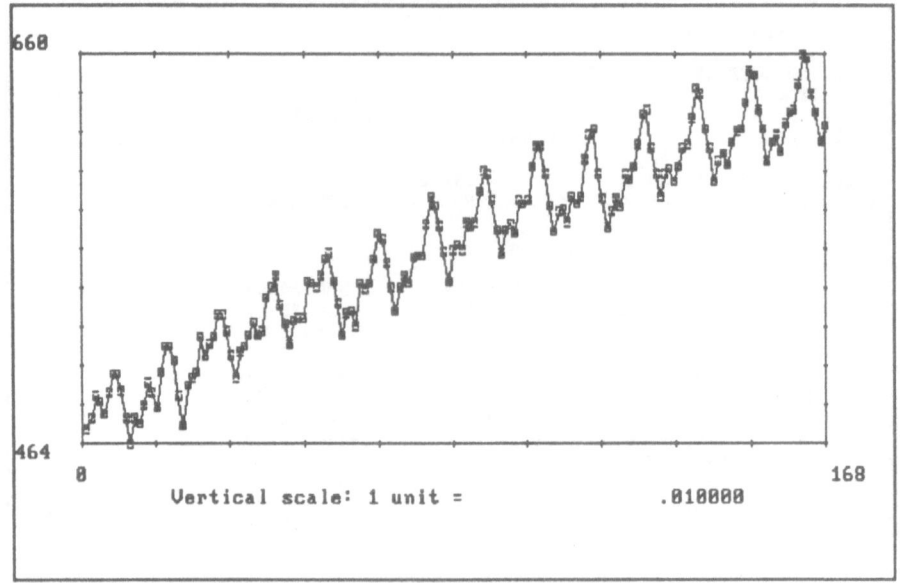

FIGURE 2.23. *The forecast values with differencing inverted*

To undo the Box-Cox transformation and recover the original data and predictors, type ↩ 0↩ y ↩ and a graph of the original AIRPASS data will be plotted on the screen. Type ↩ and the 24 predicted values will be added, giving the graph shown in Figure 2.24.

2.6   Model Properties

PEST can be used to analyze the properties of a specified ARMA process without reference to any data set. This enables us in particular to compare the properties of potential ARMA models for a given data set in order to see which of them best reproduces particular features of the data.

PEST allows you to look at the autocorrelation function and spectral density, to examine $MA(\infty)$ and $AR(\infty)$ representations and to generate realizations for any specified ARMA process. The use of these options is described in this section.

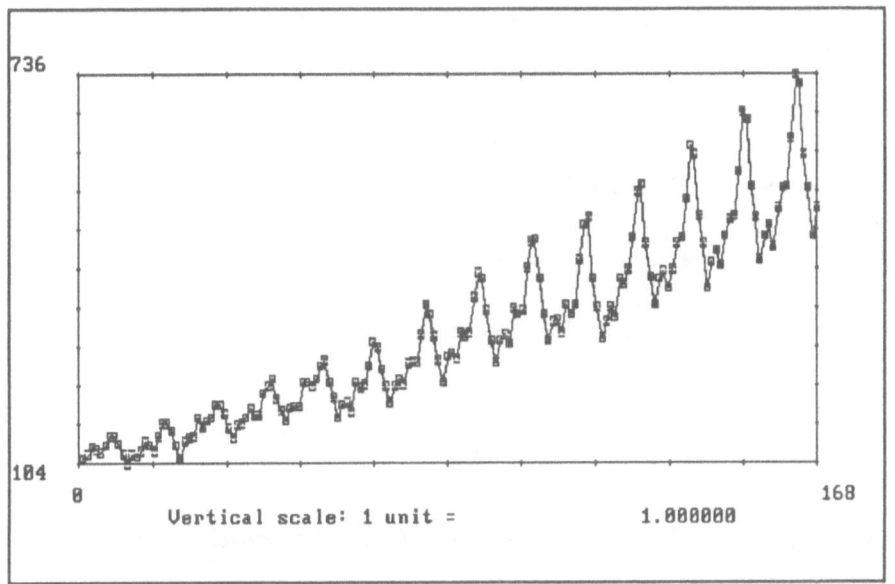

FIGURE 2.24. *The forecasts of the original AIRPASS data*

EXAMPLE: We shall illustrate the use of *PEST* for model analysis using the model AIRPASS.MOD which is currently stored in the program.

2.6.1 ARMA MODELS (*BD Chapter 3*)

For modelling zero mean stationary time series, *PEST* uses the class of ARMA processes. The initials stand for **AutoRegressive Moving Average**. *PEST* enables you to compute characteristics of specific ARMA models and to find appropriate models for given data sets (assuming of course that the data can be reasonably represented by such a model — preliminary transformations of the data may be necessary to ensure this).

$\{X_t\}$ is an ARMA(p, q) process with coefficients $\phi_1, \ldots, \phi_p, \theta_1, \ldots, \theta_q$ and white noise variance σ^2 if it is a stationary solution of the difference equations,

$$X_t = \phi_1 X_{t-1} + \phi_2 X_{t-2} + \cdots + \phi_p X_{t-p} + Z_t + \theta_1 Z_{t-1} + \theta_2 Z_{t-2} + \cdots + \theta_q Z_{t-q},$$

where $\{Z_t\} \sim \mathrm{WN}(0, \sigma^2)$ (i.e. $\{Z_t\}$ is an uncorrelated sequence of random variables with mean 0 and variance σ^2, known as a **white-noise sequence**.)

If $p = 0$ we call X_t an MA(q) (moving average of order q) process. In this case,

$$X_t = Z_t + \theta_1 Z_{t-1} + \theta_2 Z_{t-2} + \cdots + \theta_q Z_{t-q}.$$

If $q = 0$ we call X_t an AR(p) (autoregressive of order p) process. In this case,

$$X_t = Z_t + \phi_1 X_{t-1} + \phi_2 X_{t-2} + \cdots + \phi_p X_{t-p}.$$

An ARMA model is said to be **causal** if X_t has the MA(∞) representation in terms of $\{Z_t\}$,

$$X_t = \sum_{j=0}^{\infty} \psi_j Z_{t-j}, \quad t = 0, \pm 1, \pm 2, \ldots,$$

where $\sum_{j=0}^{\infty} |\psi_j| < \infty$ and $\psi_0 := 1$. If the **AR polynomial**, $1 - \phi_1 z - \cdots - \phi_p z^p$, and the **MA polynomial**, $1 + \theta_1 z + \cdots + \theta_p z^q$, have no common zeroes, then a necessary and sufficient condition for causality is that the autoregressive polynomial has no zeroes inside or on the unit circle.

PEST works exclusively with causal ARMA models. It will not permit you to enter a model for which $1 - \phi_1 z - \cdots - \phi_p z^p$ has a zero inside or on the unit circle, nor does it generate fitted models with this property. From the point of view of second order properties this represents no loss of generality (*BD Section 3.1*). If you are trying to enter an ARMA(p, q) model manually, the simplest way to ensure that your model is causal is to set all the autoregressive coefficients close to zero (e.g. .001). *PEST* will not accept a non-causal model.

An ARMA model is said to be **invertible** if Z_t can be written as

$$Z_t = \sum_{j=0}^{\infty} \pi_j X_{t-j}, \quad t = 0, \pm 1, \pm 2, \ldots,$$

where $\sum_{j=0}^{\infty} |\pi_j| < \infty$ and $\pi_0 := 1$. This condition ensures that Z_t, the noise at time t, is determined by the observations at times up to and including t, or equivalently that $\{X_t\}$ has an "AR(∞)" representation in terms of $\{Z_t\}$.

PEST does not restrict models to be invertible, however if the current model is non-invertible, i.e. if the moving average polynomial, $1 + \theta_1 z + \cdots + \theta_q z^q$ has a zero inside or on the unit circle, you will be informed by the program. (You can check the model status by choosing Option 10 of the Main Menu.) A non-invertible model can always be converted to an invertible model with the same autocovariance function by choosing Option

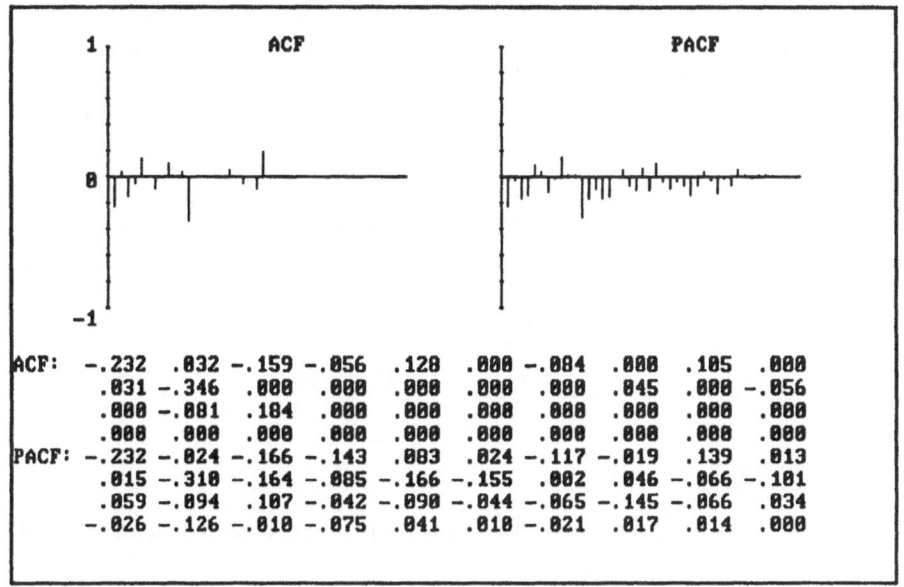

FIGURE 2.25. *The ACF and PACF of AIRPASS.MOD*

8 of the Main Menu, then Option 0 of the Estimation Menu, then Option 7 of the Results Menu.

2.6.2 MODEL ACF, PACF (*BD Sections 3.3, 3.4*)

See Section 2.3.1 for a definition of the ACF and PACF and the use of the sample ACF and PACF in model fitting.

The *model* ACF and PACF can be obtained using Option 4 of the Main Menu. They can be calculated for lags up to 1150. Normally you should not need more than about 40.

When you have specified the maximum lag, the model ACF/PACF Menu will appear. This allows you to plot the ACF and PACF, to file the values, to list the values on the screen and to change the white noise variance.

> EXAMPLE: Find the ACF and PACF for the current model AIRPASS.MOD. The results are displayed in Figure 2.25.
>
> Compare these graphs with the sample ACF and PACF shown in Figure 2.9. The data and the model ACF and PACF all have

large values at lag 12. The sample and model partial autocorrelation functions both tend to die away geometrically after the peak at lag 12. The similarities between the graphs indicate that the model is capturing some of the important features of the data.

2.6.3 MODEL REPRESENTATIONS (*BD Sections 3.1, 3.2*)

As indicated in Section 2.6.1, if $\{X_t\}$ is a causal ARMA process, then it has as an MA(∞) representation,

$$X_t = \sum_{j=0}^{\infty} \psi_j Z_{t-j}, \quad t = 0, \pm 1, \pm 2, \ldots,$$

where $\sum_{j=0}^{\infty} |\psi_j| < \infty$ and $\psi_0 = 1$.

Similarly, if $\{X_t\}$ is an invertible ARMA process, then it has an AR(∞) representation,

$$Z_t = \sum_{j=0}^{\infty} \pi_j X_{t-j}, \quad t = 0, \pm 1, \pm 2, \ldots,$$

where $\sum_{j=0}^{\infty} |\pi_j| < \infty$ and $\pi_0 = 1$.

For any specified ARMA model you can determine the coefficients in these representations by selecting Option 4 of the Main Menu. After computing the specified number of autocorrelations as described in Section 2.6.2 (you can specify 0 if you are not interested in the autocorrelations), you will see the model ACF/PACF Menu. Select Option 5 and then specify whether you require the MA(∞) or AR(∞) representation and the number of coefficients required. (If the model is not invertible the AR(∞) choice will not be possible.) *PEST* will then print the required coefficients (ψ_j or π_j) on the screen. They can be stored either as they appear on the screen or in model format for later use in *PEST*.

> EXAMPLE: AIRPASS.MOD does not have an AR(∞) representation since it is not invertible. However, we can convert AIRPASS.MOD to an equivalent invertible model and then find an AR(∞) representation for it. To convert to an invertible model, start from the Main Menu and type 8↩ 0↩ 7↩ 8↩ . To find the AR(∞) representation, type 4↩ 0↩ 5↩ 2↩ 50↩ . This gives 50 coefficients, the first 21 of which are shown in Figure 2.26.
>
> There is little point in using *PEST* to find the MA(∞) representation of this model. What is it?

```
AR-infinity coeffs up to lag    50
     j              pi (j)
     0            1.0000000
     1             .3635672
     2             .1176793
     3             .3047080
     4             .2748821
     5            -.0063462
     6             .0544923
     7             .1699656
     8             .1007748
     9             .0230251
    10             .0826018
    11             .0652756
    12             .5837110
    13             .4227796
    14             .2301330
    15             .3724042
    16             .3963610
    17             .1039688
    18             .0866742
    19             .2337619
    20             .1903337
    21             .0821539
<Press any key to continue>
```

FIGURE 2.26. *The AR(∞) representation of the invertible equivalent of AIR-PASS.MOD*

2.6.4 GENERATING REALIZATIONS OF A RANDOM SERIES (*BD Problem 8.17*)

PEST can be used to generate realizations of a random time series defined by the currently stored model.

To generate such a realization, select Option 6 from the Main Menu. You will be asked if you wish to alter the model and then prompted for the number of data points you wish to generate. Finally, *PEST* will ask you for a random number seed. This should be an integer with fewer than 10 digits. By using the same random number seed you can reproduce the same realization of the process at any other time.

Once the values of an ARMA process have been generated, you will be given the opportunity to add any specified mean to the observations. If you have previously mean-corrected a data set, the subtracted mean will have been stored by *PEST* and it will be displayed so that you may choose to add this value to the simulated ARMA data. (If you have previously performed a classical decomposition on a data set you will also be given the

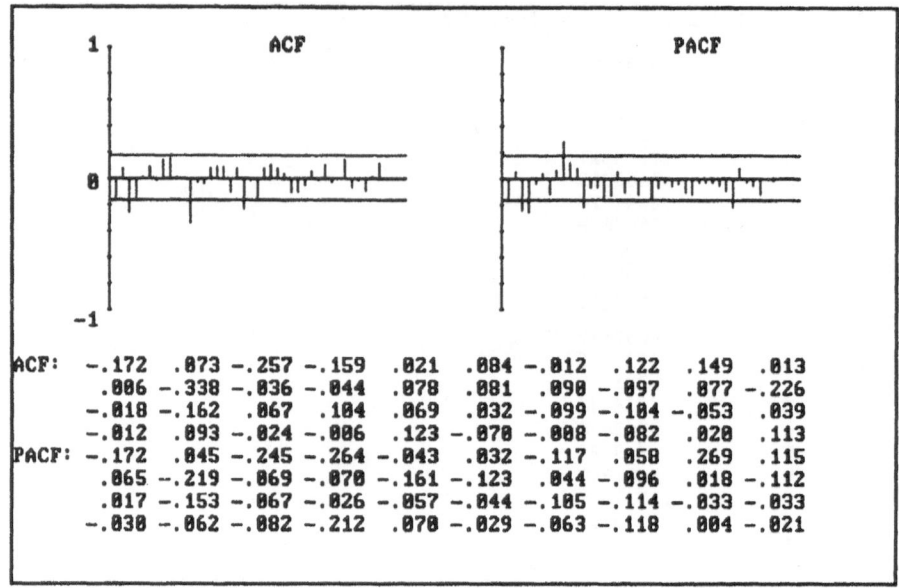

ACF: −.172 .073 −.257 −.159 .021 .084 −.012 .122 .149 .013
       .006 −.338 −.036 −.044 .070 .081 .090 −.097 .077 −.226
      −.018 −.162 .067 .104 .069 .032 −.099 −.104 −.053 .039
      −.012 .093 −.024 −.006 .123 −.070 −.008 −.082 .020 .113
PACF: −.172 .045 −.245 −.264 −.043 .032 −.117 .058 .269 .115
       .065 −.219 −.069 −.070 −.161 −.123 .044 −.096 .018 −.112
       .017 −.153 −.067 −.026 −.057 −.044 −.105 −.114 −.033 −.033
      −.030 −.062 −.082 −.212 .070 −.029 −.063 −.118 .004 −.021

FIGURE 2.27. *The sample ACF and PACF of the generated data*

opportunity to add the stored trend and seasonal components. This allows you to simulate the *original data*, not just the random noise component. If, however, you transform your original data by differencing, *PEST* allows you to simulate the differenced data only.)

The simulated data will be stored in *PEST*, overwriting any data previously stored in the program.

> EXAMPLE: To generate 135 data points using the model AIR-PASS.MOD, choose Option 6 of the Main Menu. Then type
>
> y n 135↩ 1327↩ 0↩ n
>
> (The number 1327 is the random number seed.) Print the sample ACF and PACF of the generated data using Option 3 of the Data Menu (see Figure 2.27). Compare the graphs with those in Figure 2.25. By computing the sample ACF and PACF for a variety of different realizations you can get a feeling for the magnitude of the random fluctuations in these functions.
>
> The sample ACF and PACF of the transformed airline passenger data (Figure 2.9) look equally compatible with the model

```
This procedure will find, plot and store the
spectral density of the current ARMA model
at the non-negative frequencies, J*pi/n, i.e.
        f(j*pi/n)  J = 0,...,n
where f(.) is the spectral density on (-pi,pi).

THE ARMA( 0,25) MODEL IS  X(t) = Z(t)
 +(  -.439)*Z(t- 1)  +(  -.302)*Z(t- 3)  +(   .241)*Z(t- 5)  +(  -.656)*Z(t-12)
 +(   .348)*Z(t-23)

SPECTRAL DENSITY MENU :
   1. Plot the spectral density
   2. File the spectral density
   3. Plot ln(spectral density)
   4. File ln(spectral density)
   5. Specify a diff. number of positive frequencies, now     300
   6. Change current white noise variance of     .102778E-02
   7. Return to main menu
CHOOSE A NUMBER:
```

FIGURE 2.28. *The Spectral Density Menu*

ACF and PACF (Figure 2.25) as the sample ACF and PACF
of the simulated series. This reinforces our earlier decision that
the model provides a good representation of the data.

2.6.5   MODEL SPECTRAL DENSITY (*BD Sections 4.1–4.4*)

Just as we compared the sample ACF and PACF of the data with the ACF
and PACF of the fitted model, we can compare the estimated spectral
density based on the data with the spectral density of the model. Spectral
density estimation is treated in Section 2.7. Here we consider only the
spectral density of the *model*. This is determined using Option 5 of the
Main Menu. The Spectral Density Menu is shown in Figure 2.28.

The spectral density of a stationary time series $\{X_t,\ t = 0 \pm 1, \cdots\}$ with
absolutely summable autocovariances (in particular of an ARMA process)
can be written as

$$f(\omega) = \frac{1}{2\pi} \sum_{k=-\infty}^{\infty} \gamma(k)e^{-i\omega k},\ -\pi \leq \omega \leq \pi,$$

where $\gamma(k)$ is the autocovariance at lag k and $i = \sqrt{-1}$.

The spectral representation of X_t decomposes the sequence into sinusoidal components and $f(\omega)$ measures the relative contributions to the variance of X_t from the components of different frequencies (measured in radians per unit time). For real-valued series $f(\omega) = f(-\omega)$ so it is necessary only to plot $f(\omega)$, $0 \le \omega \le \pi$. A peak in the spectral density function at frequency λ indicates a relatively large contribution to the variance from frequencies near λ.

For example the maximum likelihood AR(2) model,

$$X_t = 1.407X_{t-1} - 0.713X_{t-2} + Z_t,$$

for the data file SUNSPOTS has a peak in the spectral density at frequency $.18\pi$ radians per year. This indicates that a relatively large part of the variance of the series can be attributed to sinusoidal components with period close to $2\pi/.18 = 11.2$ years.

> EXAMPLE: Determine the spectral density of AIRPASS.MOD. Starting from the Main Menu type 5↩ 1↩ . You will then see the graph displayed in Figure 2.29.
>
> A notable feature of this graph is that there are small values at each integer multiple of $\pi/6$. These are due to our earlier differencing at lag 12 which had the effect of removing period 12 components from the data.

The other options in the Spectral Density Menu allow you to file the spectral density, and to plot and file the logarithm of the spectral density (Options 2, 3 and 4). You can change the resolution of the spectral density graph using Option 5. Finally, with Option 6 you can change the white noise variance of the model. Option 7 will take you back to the Main Menu.

2.7   Nonparametric Spectral Estimation (BD Chapter 10)

Spectral analysis is typically concerned with two problems: the detection of cyclical behavior in the data and the estimation of the spectral density. Both of these problems may be addressed by selecting Option 7 (Nonparametric spectral estimation) from the Main Menu of *PEST*. After choosing this option, the Spectral Analysis Menu (see Figure 2.30) will appear on the screen.

2.7.1   PLOTTING THE PERIODOGRAM

The periodogram and/or ln(periodogram) may be plotted by choosing Option 1 (Plot periodogram/(2*pi) and/or ln[periodogram/(2*pi)]) of the menu.

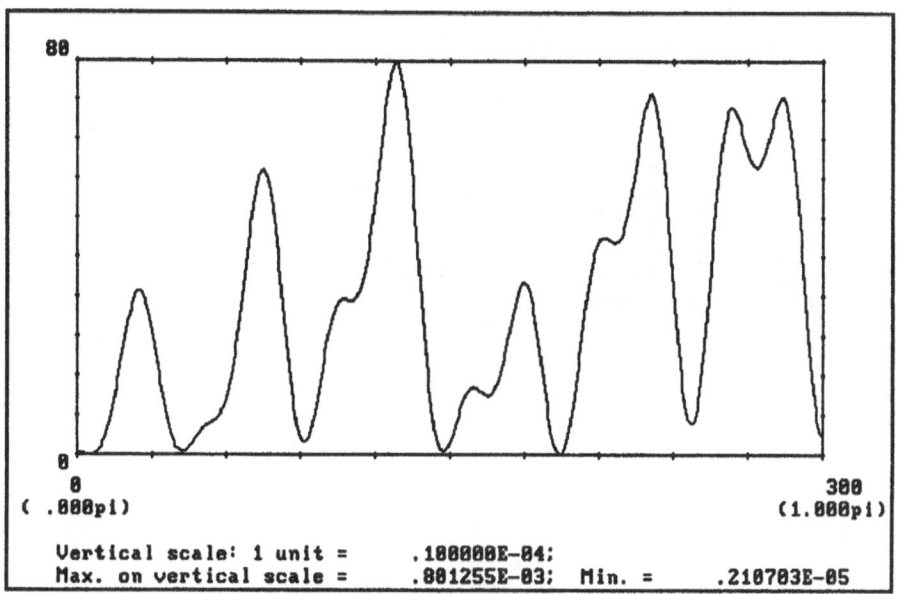

FIGURE 2.29. *The spectral density of AIRPASS.MOD*

The periodogram is defined by

$$I(\omega_j) = n^{-1} | \sum_{t=1}^{n} X_t e^{-it\omega_j} |^2$$

where $\omega_j = 2\pi j/n$, $j = 0, 1, \ldots, [n/2]$ are the Fourier frequencies in $[0, \pi]$ and $[n/2]$ is the integer part of $n/2$. (The program actually plots $I/(2\pi)$.) A large value of $I(\omega_j)$ suggests the presence of a sinusoidal component in the data at frequency ω_j. The presence of such a component may be tested using an analysis of variance table as described in *BD Section 10.1*. Alternatively, one can test for hidden periodicities in the data using the Kolmogorov–Smirnov test or Fisher's test as described below. The periodogram is computed for nonzero Fourier frequencies only, since the value at 0, $I(0) = n|\bar{X}_n|^2$, depends on the sample mean only and is generally not a useful quantity. The periodogram is computed using the fast Fourier transform. The discrete Fourier transform of the data, defined by

$$a_j = n^{-1/2} \sum_{t=1}^{n} X_t e^{-it\omega_j}, \quad -[(n-1)/2] \le j \le [n/2],$$

```
Number of observations=    131

<Computing the Fourier transform>


SPECTRAL ANALYSIS MENU :
  1. Plot periodogram/(2*pi) and/or ln[periodogram/(2*pi)]
  2. Plot cumulative periodogram
  3. File Fourier transform
  4. Fisher's test
  5. Enter weight function for smoothing
  8. Return to main menu
CHOOSE A NUMBER :
```

FIGURE 2.30. *The Spectral Analysis Menu*

may be filed using Option 3 (File Fourier transform). This option will save the coefficients $\{a_j, j = 0, \ldots, [n/2]\}$ as an array of complex numbers.

EXAMPLE: Read in the stored data file AIRPASS.LOG, difference at lags 12 and 1 as before, and subtract the mean. Then use Option 7 of the Main Menu to plot the periodogram shown in Figure 2.31. Notice the similarity between this graph and the model spectral density plotted in Figure 2.29. Now return to the Main Menu and read in the residuals AIRPASS.RES which we filed earlier after fitting AIRPASS.MOD to the transformed AIRPASS.LOG series. If we compute the periodogram of these residuals, we see (Figure 2.32) that there are no dominant frequency components (so that in this respect the residual series resembles a realization of white noise). For an iid sequence with variance σ^2 the periodogram ordinates should be approximately iid exponential variables with mean σ^2.

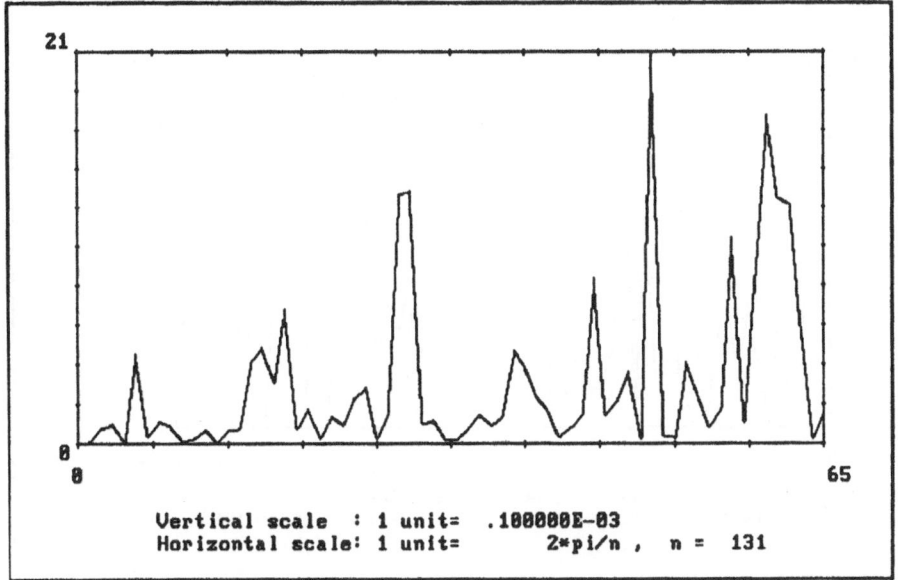

FIGURE 2.31. *Periodogram of the twice differenced AIRPASS.LOG series*

2.7.2   PLOTTING THE CUMULATIVE PERIODOGRAM

Select Option 2 (Plot cumulative periodogram) of the Spectral Analysis Menu to plot the standardized cumulative periodogram defined as

$$C(x) = \begin{cases} 0, & x < 1 \\ Y_i, & i \le x < i+1, \ i = 1, \ldots, q-1, \\ 1, & x \ge q, \end{cases}$$

where $q = [(n-1)/2]$ and

$$Y_i = \frac{\sum_{k=1}^{i} I(\omega_k)}{\sum_{k=1}^{q} I(\omega_k)}.$$

If $\{X_t\}$ is Gaussian white noise, then $Y_i, i = 1, \ldots, q-1$ are distributed as the order statistics from a sample of $q - 1$ independent uniform(0,1) random variables, and the standardized cumulative periodogram should be approximately linear. The hypothesis of Gaussian white noise is rejected

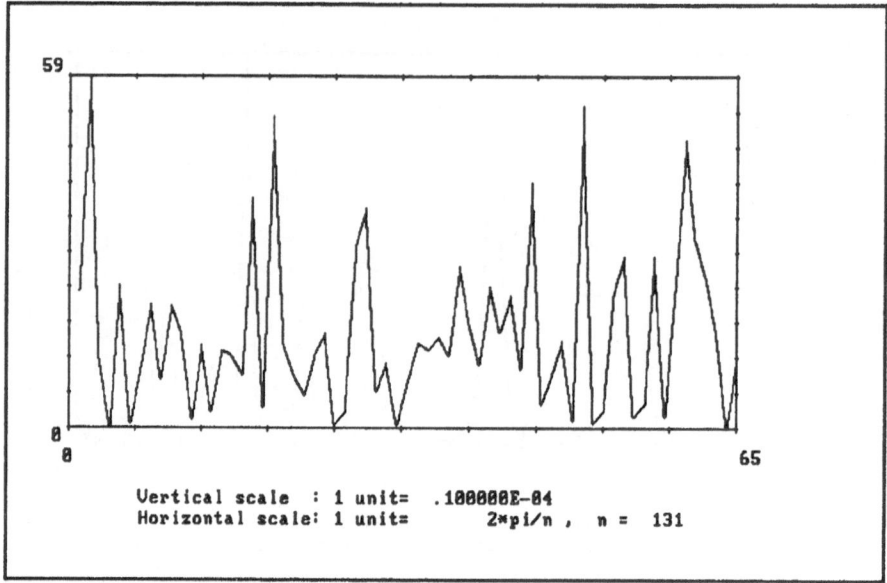

FIGURE 2.32. *Periodogram of the residuals from AIRPASS.MOD*

at level .05 if $C(x)$ exits from the boundaries

$$y = \frac{x-1}{q-1} \pm 1.36(q-1)^{-1/2}, \quad 1 \le x \le q.$$

EXAMPLE: After returning to the Spectral Analysis Menu, type
2↵ to plot the standardized cumulative periodogram (Figure
2.33) for AIRPASS.RES. The function $C(x)$ lies well within the
above boundaries (here $q = [(131-1)/2] = 65$), supporting the
hypothesis that the residuals are observations of independent
white noise.

2.7.3   FISHER'S TEST

Fisher's test enables you to test the data for the presence of hidden peri-
odicities with unspecified frequency. If the test statistic defined by

$$\xi_q = \frac{\max_{1 \le i \le q} I(\omega_i)}{q^{-1} \sum_{i=1}^{q} I(\omega_i)}$$

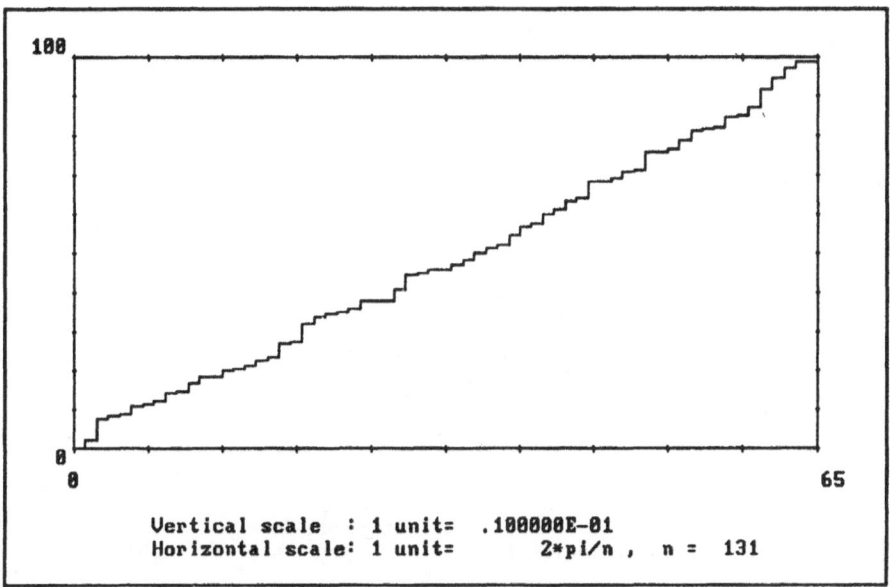

FIGURE 2.33. *Cumulative periodogram of the residuals*

is large then the hypothesis that the data is Gaussian white noise is rejected. Option 4 (Fisher's test) of the Spectral Analysis Menu gives the observed value of ξ_q and the p-value of the test (i.e. the probability that ξ_q exceeds the observed value under the null hypothesis that the data is Gaussian white noise).

> EXAMPLE: To apply Fisher's test to AÏRPASS.RES, choose Option 4 of the Spectral Analysis Menu and you will see the display,

```
Observed ratio of maximum periodogram to average  =    3.5954
Probability (under Ho) of ratio larger than observed = 0.8840
```

> Since the p-value is rather large, the assumption of independence of the residuals is not incompatible with Fisher's test.

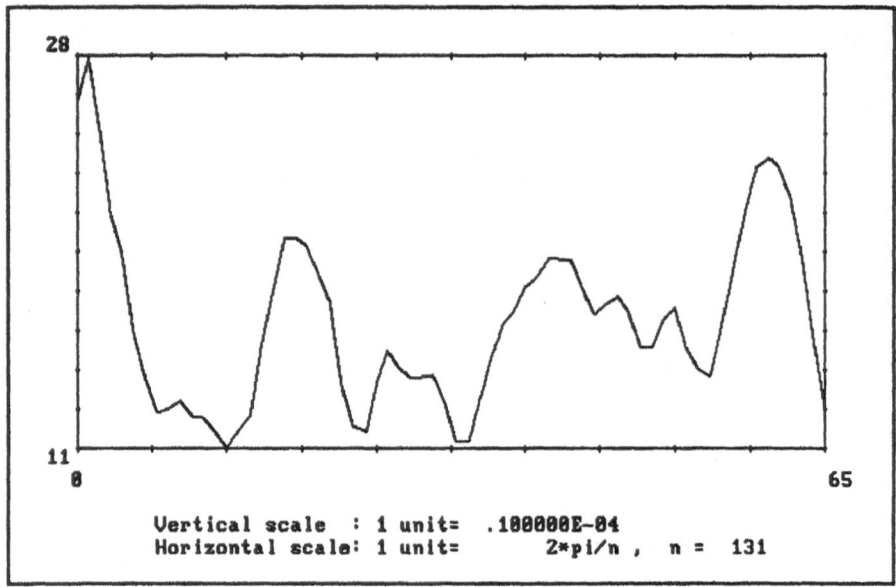

FIGURE 2.34. *Smoothed spectrum estimate for AIRPASS.RES*

2.7.4   SMOOTHING TO ESTIMATE THE SPECTRAL DENSITY (*BD Section 10.4*)

The spectral density of a stationary process is estimated by smoothing the periodogram. The weight function $\{W(j), |j| \leq m\}$ used for smoothing is entered through Option 5 (Enter weight function for smoothing) of the menu. After typing 5↩ , you will be asked to enter a value for m. Type -1↩ if a weight function is to be read from a file and type 0↩ if you want to return to the Spectral Analysis Menu. For positive values of m you will be requested to enter the weights $W(0), W(1), \ldots, W(m)$, all of which must be nonnegative. The program ensures that the weight function is symmetric by defining $W(-j) = W(j), j = 1, \ldots, m$, and then rescales the weights so that they add to one. After the weights have been entered, the program returns to the Spectral Analysis Menu which now contains the 2 new items, 6. Plot weight function and 7. Plot estimated spectrum and/or ln[spectrum].

EXAMPLE: Try estimating the spectral density of the data file AIRPASS.RES using the weight function, $W(0) = W(1) = W(2) = \frac{3}{21}, W(3) = \frac{2}{21}$, and $W(4) = \frac{1}{21}$. Begin by typing

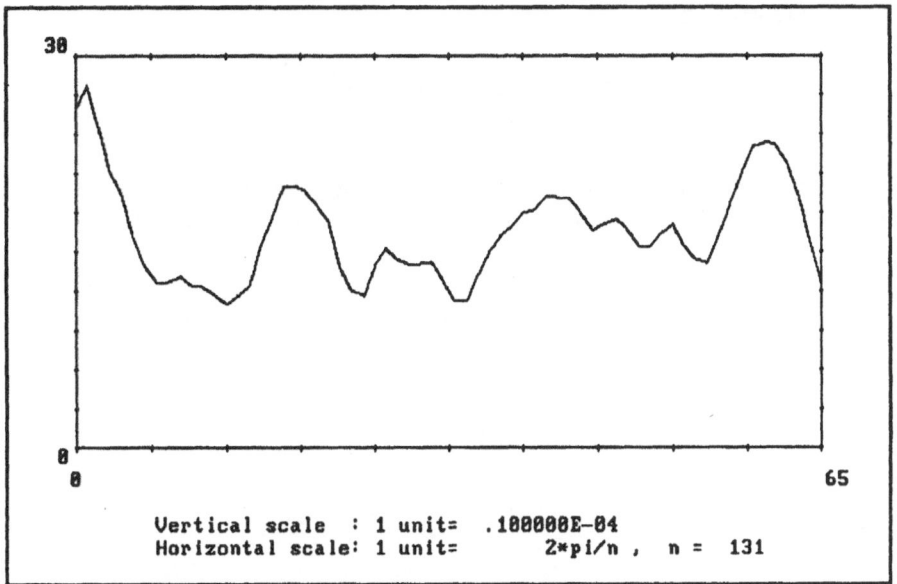

FIGURE 2.35. *Rescaled spectrum estimate for AIRPASS.RES*

5↩ 4↩ 3↩ 3↩ 3↩ 2↩ 1↩ . (The program automatically
divides the weights entered by 21 so that they add to 1). Plot
the weight function by typing 6↩ . The entries ↩ 1↩ n return you to the Spectral Analysis Menu. Type 7↩ to plot the
smoothed periodogram (Figure 2.34). This can be plotted on a
more natural scale by typing ↩ 1↩ .0003↩ 0↩ (see Figure
2.35). Approximate 95% confidence bounds for the true spectral
density, $f(\omega_j)$, are given (*BD Section 10.4*) by

$$\hat{f}(\omega_j) \quad \pm \quad 1.96 \left(\sum_{|k| \leq m} W^2(k) \right)^{1/2} \hat{f}(\omega_j) \quad \text{or}$$
$$\hat{f}(\omega_j) \quad \pm \quad .6921 \hat{f}(\omega_j).$$

These bounds are compatible with the constant spectral density
of white noise.

The estimate $\ln \hat{f}$ of the ln[spectrum] can be plotted by typing
↩ 0↩ n y. Approximate 95% confidence bounds for $\ln f(\omega_j)$

are given by

$$\ln \hat{f}(\omega_j) \;\pm\; 1.96 \left(\sum_{|k| \le m} W^2(k) \right)^{1/2} \quad \text{or}$$
$$\ln \hat{f}(\omega_j) \;\pm\; .6921.$$

It is often more convenient to make inferences for $\ln f$ since the widths of the confidence intervals are the same for all frequencies.

3

SMOOTH

3.1 Introduction (*BD Section 1.4*)

To run the program *SMOOTH*, type SMOOTH↩ . After entering the appropriate graphics code number you will be asked to specify the file name of the input series, $\{X_t, t = 1, \ldots, n\}$, which is to be smoothed. The program allows you to compute, plot and file the values of the smoothed series, $\{\hat{m}_t, t = 1, \ldots, n\}$, which can be specified either as

- a symmetric moving average,

$$\hat{m}_t = \sum_{j=-q}^{q} a(j)X_{t-j}, \quad t = 1, \ldots, n,$$

  where $X_t := X_1$ for $t < 1$ and $X_t := X_n$ for $t > n$, or

- a one-sided exponentially weighted moving average defined by the recursions, $\hat{m}_1 = X_1$ and

$$\hat{m}_t = aX_t + (1 - a)\hat{m}_{t-1}, \quad t = 2, \ldots, n,$$

  where a is a specified smoothing constant $(0 \le a \le 1)$.

The choice is made from the menu which appears as below once the input file has been read by the program. There are four options:

```
MENU :
    1. Smooth the data using a symmetric moving average
    2. Apply exponential smoothing to the data
    3. Enter a new data set
    4. Exit from the program
CHOOSE A NUMBER :
```

3.2 Moving Average Smoothing

If you select Option 1 of the menu you will be asked to enter the half-length q and the coefficients $a(0), a(1), \ldots, a(q)$ of the required moving average,

$$\hat{m}_t = \sum_{j=-q}^{q} a(j)X_{t-j}, \quad t = 1, \ldots, n,$$

where $a(j) = a(-j),\ \ j = 1, \ldots, q$.

The integer q can take any any value greater than or equal to zero and less than $n/2$.

You may enter any real numbers for the coefficients $a(j), j = 0, \ldots, q$. These will automatically be rescaled by the program so that $a(0) + 2a(1) + \cdots + 2a(q) = 1$. (This is achieved by dividing each entered coefficient by the sum $a(0) + 2a(1) + \cdots + 2a(q)$. The program therefore prevents you from entering weights for which this sum is zero.)

Once the parameters $q, a(0), \ldots, a(q)$, have been entered, the program will print on the screen the square root of the average squared deviation of the smoothed values from the original observations, i.e.,

$$\text{SQRT(MSE)} = \sqrt{n^{-1} \sum_{j=1}^{n} (\hat{m}_t - X_t)^2}.$$

It will then plot the original data. Typing ↩ , will cause the smoothed values to be plotted on the same graph. When plotting has been completed you will be given the option of filing the smoothed values. Finally you will be returned to the menu and offered the four choices listed in Section 3.1.

> EXAMPLE: To smooth the data set STRIKES on Disk 1 using the moving average with weights $a(j) = .2,\ j = -2, -1, 0, 1, 2$, and $a(j) = 0,\ |j| > 2$, use the following sequence of entries (for graphics we assume you have an EGA card) : SMOOTH ↩ 3↩ 0↩ STRIKES ↩ ↩ 1↩ ↩ 2↩ 1↩ 1↩ 1↩ ↩ .
>
> At this point the screen will display the value
> SQRT(MSE)=1956.142000
>
> Typing ↩ ↩ will then produce the graph displayed in Figure 3.1. The points denoted by squares are the original data and the points joined by a continuous line are the smoothed values.
>
> To file the smoothed values under the file name SMST, type ↩ 0↩ y SMST ↩ . The smoothed values will then be filed under the name specified and the screen will again display the menu listed above in Section 3.1.

3.3 Exponential Smoothing

If you select Option 2 of the menu you will be asked to enter the parameter a in the smoothing recursions, $\hat{m}_1 = X_1$ and

$$\hat{m}_t = aX_t + (1 - a)\hat{m}_{t-1},\ \ t = 2, \ldots, n.$$

The choice $a = 1$ gives no smoothing and the choice $a = 0$ gives $\hat{m}_t = X_1,\ t = 1, \ldots, n$. Any value for a between these two extremes can be chosen.

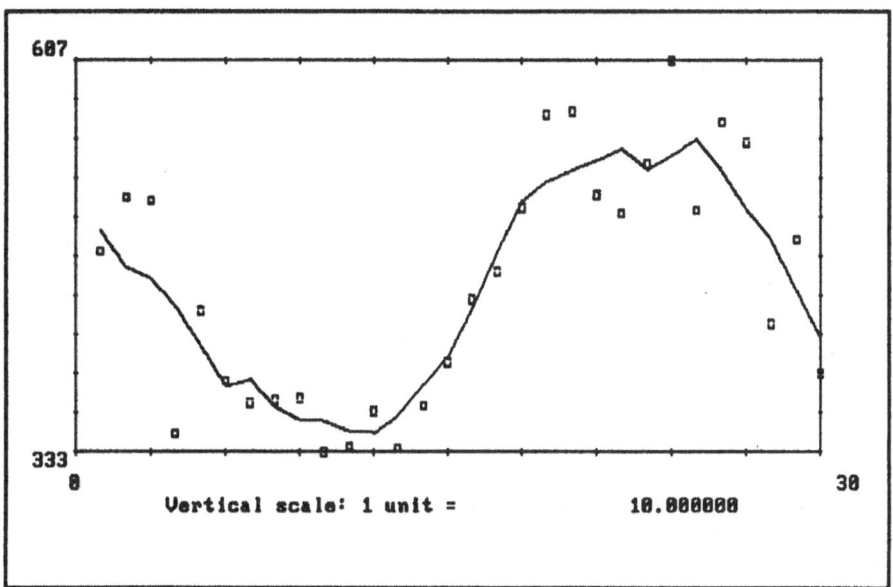

FIGURE 3.1. *The STRIKES data with smoothed values obtained from a simple moving average of length 5*

Once the parameter a has been entered, the program will print on the screen the square root of the average squared deviation of the smoothed values from the original observations, i.e.,

$$\text{SQRT(MSE)} = \sqrt{n^{-1} \sum_{j=1}^{n} (\hat{m}_t - X_t)^2}.$$

It will then plot the original data. Typing ↩ will cause the smoothed values to be plotted on the same graph. When plotting has been completed you will be given the option of filing the smoothed values. Finally you will be returned to the menu and offered the four choices listed in Section 3.1.

EXAMPLE: Continuing with the example in Section 3.2 from the point where the menu was displayed on the screen, we can now exponentially smooth the data set STRIKES using the following entries (we shall take $a = .4$) : 2↩ ↩ .4↩

At this point the screen will display the value
SQRT(MSE)=1965.049000

Typing ↩ ↩ will then produce the graph displayed in Figure

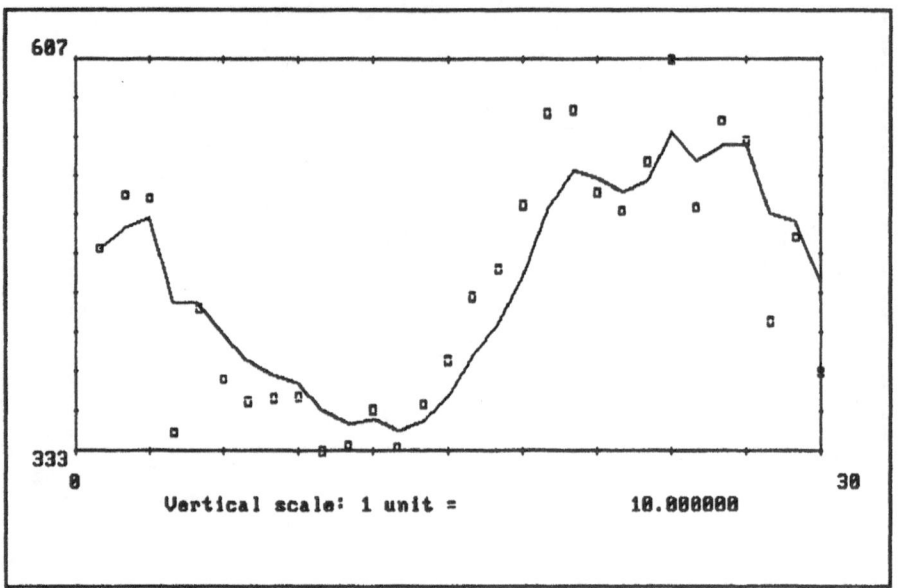

FIGURE 3.2. *The STRIKES data showing smoothed values obtained by exponential smoothing with parameter a = .4*

3.2. The points denoted by squares are the original data and the points joined by a continuous line are the smoothed values.

To file the smoothed values under the file name SMST, type ↵ 0↵ y SMST ↵ . The smoothed values will then be filed under the name specified and the screen will again display the menu listed in Section 3.1.

4

SPEC

4.1 Introduction

To run the program *SPEC*, type SPEC↩ . After entering the appropriate graphics code number, the following question will appear on your screen:

Spectral Analysis for 1 or 2 data sets?

Type 1↩ for univariate spectral analysis or 2 ↩ for bivariate spectral analysis. If a univariate analysis is desired, then, after entering the name of the input series, a menu will appear which is practically identical to the Spectral Analysis Menu (Option 7 (Nonparametric spectral estimation (SPEC)) of the Main Menu of *PEST*). The reader is referred to Section 2.7 for instructions on using this portion of *SPEC*.

4.2 Bivariate Spectral Analysis (*BD Section 11.7*)

If a bivariate analysis is requested, then you will be asked to specify the file names of the first $\{X_{t1}, t = 1, \ldots, n\}$ and second $\{X_{t2}, t = 1, \ldots, n\}$ time series. (Each series must be stored in separate ASCII files.) Once the 2 series have been successfully read by the program, the Bivariate Spectral Analysis Menu will appear with the following options:

```
1. Plot estimated f11 (no smoothing) for series 1
2. Plot estimated f22 (no smoothing) for series 2
3. Plot estimated abs. coherency |K12| (no smoothing)
4. Plot estimated phase spectrum PHI12 (no smoothing)
5. Enter weight function for smoothing
7. Begin another analysis
8. End program
```

At this point a weight function has not yet been specified, so that the estimated spectral densities of the first and second series, $\hat{f}_{11}$ and $\hat{f}_{22}$, are just the respective periodograms divided by 2π, i.e.,

$$
\begin{aligned}
\tfrac{1}{2\pi} I_{11}(\omega_k) &= \tfrac{1}{2\pi} n^{-1} \left| \sum_{t=1}^{n} X_{t1} e^{-it\omega_k} \right|^2, \\
\tfrac{1}{2\pi} I_{22}(\omega_k) &= \tfrac{1}{2\pi} n^{-1} \left| \sum_{t=1}^{n} X_{t2} e^{-it\omega_k} \right|^2,
\end{aligned}
$$

where $\omega_k = 2\pi k/n, k = 0, 1, \ldots, [n/2]$ are the Fourier frequencies in $[0, \pi]$ and $[n/2] =$ integer part of $n/2$. The cross periodogram is defined by

$$I_{12}(\omega_k) = n^{-1}\left(\sum_{t=1}^{n} X_{t1}e^{-it\omega_k}\right)\overline{\left(\sum_{t=1}^{n} X_{t2}e^{-it\omega_k}\right)}.$$

Without smoothing, the estimated absolute coherency spectrum is

$$|\mathcal{K}_{12}(\omega_k)| = \frac{|I_{12}(\omega_k)|}{I_{11}^{1/2}(\omega_k)I_{22}^{1/2}(\omega_k)} = 1.$$

In order to find a meaningful estimate of the absolute coherency spectrum (and better estimates of the marginal and phase spectra), it is necessary to smooth the periodogram.

The weight function $\{W(j), |j| \leq m\}$, for smoothing the periodogram, is entered through Option 5 (Enter weight function for smoothing) of the Bivariate Spectral Analysis Menu. After typing 5 ↩ you will be asked to enter a value for m. Type -1 ↩ if a weight function is to be read from a file and 0 ↩ if you wish to return to the Bivariate Spectral Analysis Menu without entering a weight function. For positive values of m you will be requested to enter the weights $W(0), W(1), \ldots, W(m)$, all of which must be nonnegative. The program ensures that the weight function is symmetric by defining $W(-j) = W(j), j = 1, \ldots, m$ and then rescales the weights so that they add to one. After the weights have been entered, the program returns to the menu which now contains the new item: **6. Plot weight function**.

4.2.1 ESTIMATING THE SPECTRAL DENSITY OF EACH SERIES

After the weight function has been entered, the marginal spectral densities and cross spectral density are estimated as

$$\begin{aligned}
\hat{f}_{11}(\omega_j) &= \tfrac{1}{2\pi}\sum_{|k| \leq m} W(k)I_{11}(\omega_j + \omega_k), \\
\hat{f}_{22}(\omega_j) &= \tfrac{1}{2\pi}\sum_{|k| \leq m} W(k)I_{22}(\omega_j + \omega_k), \\
\hat{f}_{12}(\omega_j) &= \tfrac{1}{2\pi}\sum_{|k| \leq m} W(k)I_{12}(\omega_j + \omega_k).
\end{aligned}$$

The two marginal spectral densities may be plotted by selecting Options 1 and 2 of the Bivariate Spectral Analysis Menu.

> EXAMPLE: Use *PEST* to difference the series LEAD and SALES each once at lag 1. File the resulting series as DLEAD and DSALES respectively. We can now do a bivariate spectral analysis of DLEAD and DSALES using *SPEC*. Type SPEC↩ , enter your graphics code number as described in Section 1.2 and let

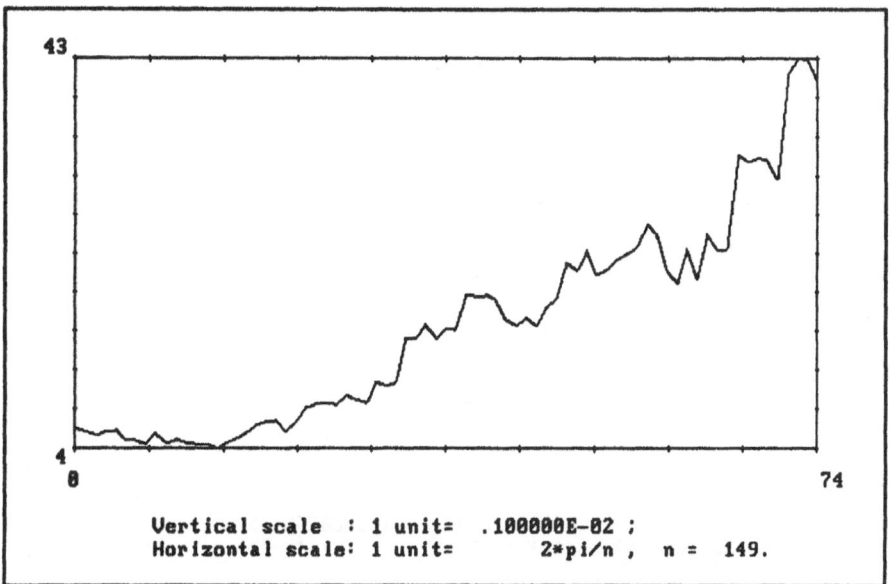

Vertical scale : 1 unit= .100000E-02 ;
Horizontal scale: 1 unit= 2*pi/n , n = 149.

FIGURE 4.1. *Smoothed spectrum estimate for DLEAD*

SPEC know that you require an analysis for a bivariate time
series. When asked for the names of the first and second time
series, type DLEAD ↩ and DSALES ↩ respectively. The Bivariate Spectral Analysis Menu should then appear on the screen.
Next try computing smoothed spectrum estimates with weight
function $W(0) = W(1) = \cdots = W(6) = \frac{1}{13}$. This is done by typing 5 ↩ 6 ↩ 1↩ 1↩ 1↩ 1↩ 1↩ 1↩ 1↩ . The program
will automatically rescale the entered weights so that they sum
to 1. Plot the weight function by typing 6 ↩ . The sequence of
entries ↩ 0 ↩ n returns you to the Main Menu. The estimated
spectral densites of DLEAD and DSALES can now be plotted
by choosing Options 1 and 2 (see Figures 4.1 and 4.2). Confidence intervals for either the spectrum of DLEAD or DSALES
are computed as described in Section 2.7.4.

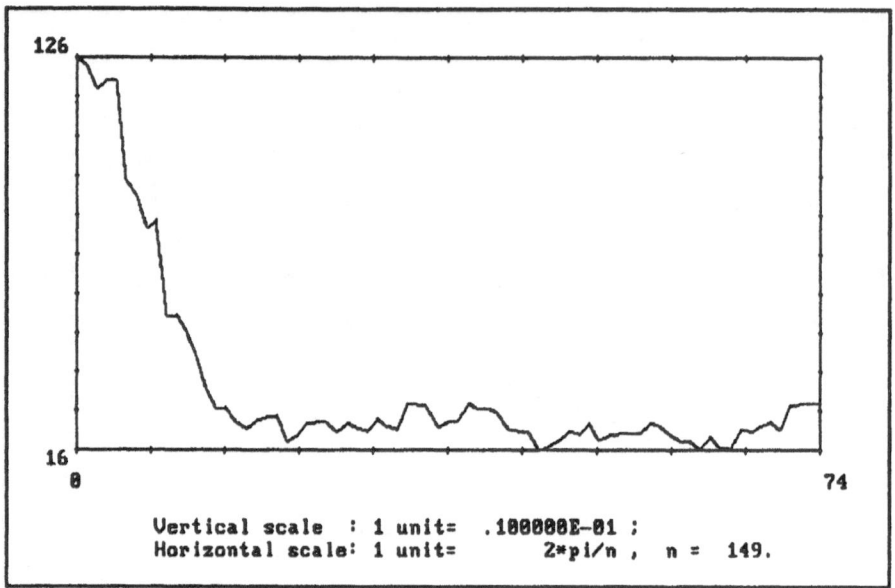

FIGURE 4.2. *Smoothed spectrum estimate for DSALES*

4.2.2 ESTIMATING THE ABSOLUTE COHERENCY SPECTRUM

The absolute coherency spectrum is estimated by

$$|\hat{\mathcal{K}}_{12}(\omega_j)| = \frac{|\hat{f}_{12}(\omega_j)|}{\hat{f}_{11}^{1/2}(\omega_j)\hat{f}_{22}^{1/2}(\omega_j)}$$

where $\hat{f}_{12}(\cdot)$ is the estimate of the cross spectrum given by

$$\hat{f}_{12}(\omega_j) = \frac{1}{2\pi}\sum_{|k|\leq m} W(k)I_{12}(\omega_j + \omega_k).$$

Roughly speaking, the absolute coherency at frequency λ is the absolute value of the correlation between the frequency-λ harmonic components in the two series (see *BD Sections 11.6 and 11.7*). An absolute coherency near 1 indicates a strong linear relationship between the sinusoidal components in the two series.

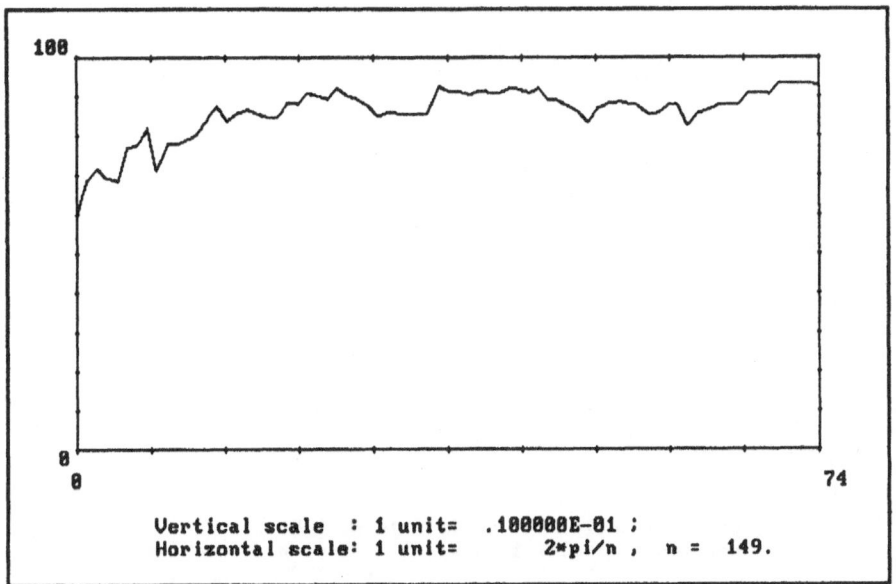

FIGURE 4.3. *Estimated absolute coherency for DLEAD–DSALES data*

EXAMPLE: For the DLEAD and DSALES data, select Option 3 from the Bivariate Spectral Analysis Menu to plot the estimated absolute coherency (see Figure 4.3). For this example, the estimated absolute coherency is rather large for all Fourier frequencies. A $100(1-\alpha)\%$ confidence interval for $|\mathcal{K}_{12}(\omega_j)|$ is the intersection of $[0,1]$ with the interval

$$(\tanh[\tanh^{-1}(\hat{\mathcal{K}}_{12}(\omega_j)) - \Phi_{1-\alpha/2}a_n/\sqrt{2}],$$
$$\tanh[\tanh^{-1}(\hat{\mathcal{K}}_{12}(\omega_j)) + \Phi_{1-\alpha/2}a_n/\sqrt{2}])$$

where $a_n^2 = \sum_{|k|\le m} W^2(k)$ and Φ_α is the α percentile of a standard normal distribution (see *BD equation (11.7.13)*). For this example, $a_n = 1/\sqrt{13}$. The lower limit of the 95% confidence interval for $|\mathcal{K}_{12}(\omega_j)|$ is bounded well away from the 0 which suggests that the absolute coherency is positive for all frequencies.

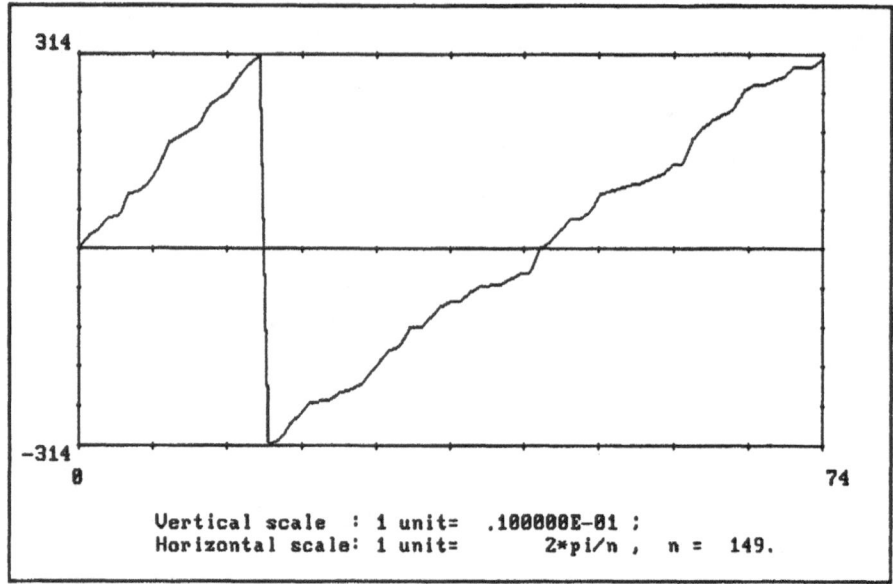

FIGURE 4.4. *Estimated phase spectrum for DLEAD–DSALES data*

4.2.3   ESTIMATING THE PHASE SPECTRUM

The phase spectrum, $\phi_{12}(\cdot) \in [-\pi, \pi]$, is defined as $\arg(f_{12}(\lambda))$. The phase spectrum is a measure of the the phase lag of the frequency-λ component of $\{X_{t2}\}$ behind that of $\{X_{t1}\}$. The derivative of $\phi_{12}(\lambda)$ can be interpreted as the time lag by which the frequency-λ component of X_{t2} follows that of X_{t1}. For example if $\phi_{12}(\lambda)$ is piecewise linear with slope d, then X_{t2} lags d time units behind X_{t1}. The phase spectrum is estimated by

$$\hat{\phi}_{12}(\omega_j) = \arg(\hat{f}_{12}(\omega_j)).$$

EXAMPLE: Continuing with the DLEAD–DSALES example, select Option 4 to plot the estimated phase spectrum (see Figure 4.4). The graph of $\hat{\phi}_{12}(\omega_j)$ is roughly piecewise linear with slope 4.1 at the low frequencies and slope 2.7 at higher frequencies. This suggests that DSALES follows DLEAD by approximately 3 time lags. A transfer function model with input DLEAD and output DSALES (see Chapter 5 and *BD Section 13.1*) and a bivariate AR model fitted to the two-dimensional series (DLEAD,DSALES) (see Chapter 6 and *BD Section 11.5*) both support this observation.

5

TRANS

5.1 Introduction (*BD Section 13.1*)

To run the program *TRANS*, type TRANS↩ . After entering the appropriate
graphics code number you will see the following menu:

```
MAIN MENU:
   1. Compute the sample cross correlations of 2 series.
   2. Fit a preliminary MA(m) transfer function model,
         X2(t) = t(0)X1(t)+...+t(m)X1(t-m)+N(t),
      to data from the zero-mean stationary series, X1, X2.
      For this option you will need files containing
      R1, the residuals obtained from PEST by fitting
      an ARMA model to the observations of X1, and
      R2, the residuals obtained from PEST when the
      same ARMA filter is applied to X2.
   3. Calculate the residuals from a specified rational
      transfer function relation between the mean corrected
      (possibly differenced) versions of two specified
      input and output series.
   4. Fit a transfer function model to a sequence of
      observations of inputs and outputs and use it to
      predict future outputs.
   5. Exit from the program.
   CHOOSE A NUMBER :
```

5.2 Computing Cross Correlations (*BD Section 11.2*)

If you select Option 1 of the Main Menu, you will be asked to enter the
file names of the first series $\{Y_1(t), t = 1, \ldots, n\}$ and the second series
$\{Y_2(t), t = 1, \ldots, n\}$.

Option 1 can then be used to calculate the sample cross correlations

$$\hat{\rho}_{Y_1,Y_2}(h) = \hat{\gamma}_{Y_1,Y_2}(h)(\hat{\gamma}_{Y_1,Y_1}(0)\hat{\gamma}_{Y_2,Y_2}(0))^{-1/2}, \ |h| < n,$$

where

$$\hat{\gamma}_{Y_i,Y_j}(h) = \begin{cases} n^{-1}\sum_{t=1}^{n-h}(Y_i(t+h) - \overline{Y}_i)(Y_j(t) - \overline{Y}_j), & h \geq 0, \\ \\ n^{-1}\sum_{t=-h+1}^{n}(Y_i(t+h) - \overline{Y}_i)(Y_j(t) - \overline{Y}_j), & h \leq 0. \end{cases}$$

Alternatively you may apply up to two differencing operators to the input series (where the same operators are applied to both series) and compute the sample cross correlations of the resulting series. For example, if you select two differencing operators with lags l_1 and l_2, the program will compute the sample cross correlations of $\{X_1(t)\}$ and $\{X_2(t)\}$, where

$$X_1(t) = (1 - B^{l_1})(1 - B^{l_2})Y_1(t), \quad t = l_1 + l_2 + 1, \ldots, n$$

and

$$X_2(t) = (1 - B^{l_1})(1 - B^{l_2})Y_2(t), \quad t = l_1 + l_2 + 1, \ldots, n.$$

EXAMPLE: To compute the sample cross correlations of the two data sets Y_1 =DATA\LEAD and Y_2 =DATA\SALES on Disk 2, use the following sequence of entries (for graphics we assume you have an EGA card):

TRANS $\hookleftarrow$ **3**$\hookleftarrow$ **0**$\hookleftarrow$ **1**$\hookleftarrow$ **DATA\LEAD** $\hookleftarrow$ **DATA\SALES** $\hookleftarrow$ **0**$\hookleftarrow$ $\hookleftarrow$

At this point the screen will display the graph of cross correlations shown in Figure 5.1.

When you have inspected the graph, type $\hookleftarrow$ and you will be asked if you wish to list the sample autocorrelations on the screen. Type y and you will see a listing of $\hat{\rho}_{Y_1,Y_2}(h), h = -30, -29, \ldots, 30$ and you will be asked whether or not you wish to file the cross correlations. Type n and you will then be returned to the Main Menu.

Inspection of the graphs of the two data sets Y_1 =DATA\LEAD and Y_2 =DATA\SALES and their autocorrelations using *PEST* suggests a single differencing at lag 1 to make the series stationary. If X_1 and X_2 denote the series

$$X_i(t) = (1 - B)Y_i(t), \quad i = 1, 2,$$

then the sample autocorrelation function of X_1 and X_2 can be computed using the following entries after the appearance on the screen of the Main Menu:

1$\hookleftarrow$ **DATA\LEAD** $\hookleftarrow$ **DATA\SALES** $\hookleftarrow$ **1**$\hookleftarrow$ **1**$\hookleftarrow$ $\hookleftarrow$
dlead$\hookleftarrow$ **dsales**$\hookleftarrow$

At this point the screen will display the graph of cross correlations shown in Figure 5.2. As before you will be given the options of listing and filing the cross correlations before being returned to the Main Menu.

5.3   An Overview of Transfer Function Modelling

- Given observations of an "input" series $\{Y_1(t)\}$ and an "output" series $\{Y_2(t)\}$, the steps in setting up a transfer function model relating Y_2

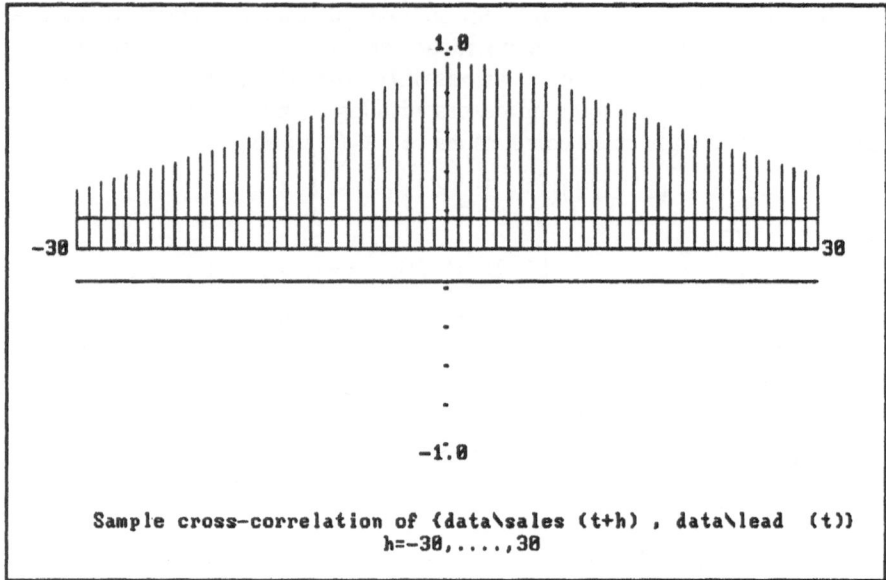

FIGURE 5.1. *The cross correlations of DATA\LEAD and DATA\SALES*

to Y_1 begin with differencing and mean correction to generate transformed input and output series X_1 and X_2 which can be modelled as zero mean stationary processes. Suitable differencing operators (up to two are allowed by *TRANS*) can be found by examination of the series Y_1 and Y_2 using *PEST*. The same differencing operations must be applied to both series.

- An ARMA model is fitted to the transformed input series X_1 using *PEST*, and the residual series R_1 is filed for later use. The same ARMA filter is then applied to X_2 using Option 0 of the Estimation Menu of *PEST* (select Option 8 from the Main Menu followed by Option 0). The residual series R_2 is then filed.

- A preliminary transfer function model relating X_2 to X_1 is found using Option 2 of *TRANS*. This has the form,

$$X_2(t) = \sum_{j=0}^{m} t(j)X_1(t - j) + N(t),$$

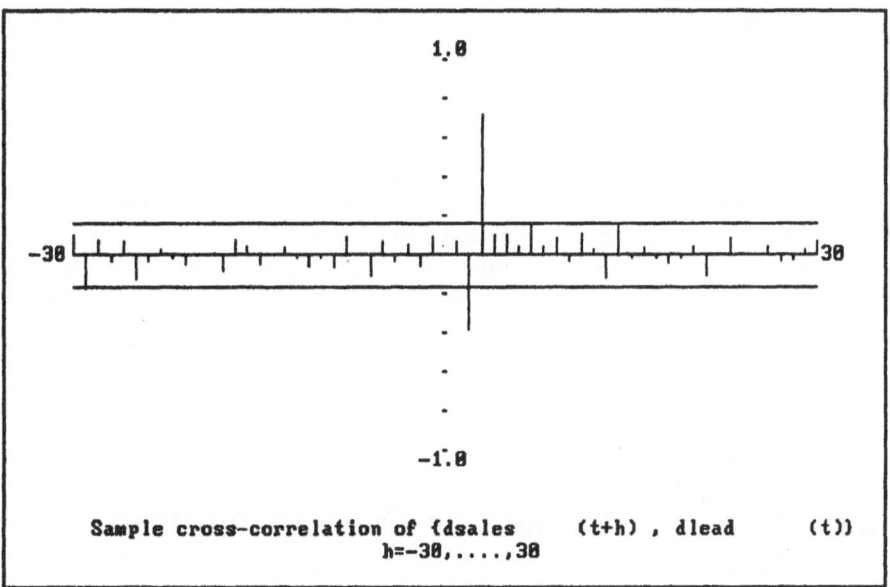

FIGURE 5.2. *The cross correlations of DLEAD and DSALES, obtained by differencing DATA\LEAD and DATA\SALES at lag 1*

where $\{N(t)\}$ is a zero mean stationary noise sequence.

- It is often convenient to replace the transfer function $\sum_{j=0}^{m} t(j)B^j$ by a rational function of B with fewer coefficients. For example, the transfer function,

$$2B + .22B^2 + .018B^3 + .002B^4,$$

could be approximated by the more *parsimonious* transfer function,

$$T(B) = \frac{2B}{1 - .1B}.$$

- Given the series X_1 and X_2 and given any rational transfer function $T(B)$, Option 3 of *TRANS* calculates values of the noise series $\{N(t)\}$ in the model

$$X_2(t) = T(B)X_1(t) + N(t).$$

- An ARMA model $\phi(B)N(t) = \theta(B)W(t)$ is then fitted to the noise series $\{N(t)\}$. This gives the preliminary transfer function model,

$$X_2(t) = T(B)X_1(t) + \phi^{-1}(B)\theta(B)W(t).$$

- Option 4 of *TRANS* requires that you enter the preliminary model just determined. It reestimates the coefficients in the preliminary model using least squares. A Kalman filter representation of the model is used to determine minimum mean squared error linear predictors of the output series. Model selection can be made with the AICC statistic, which is computed for each fitted model. Model checking can be carried out by checking the residuals for whiteness and checking the cross correlations of the input residuals and the transfer function residuals.

5.4   Fitting a Preliminary Transfer Function Model

Option 2 of the Main Menu of *TRANS* is concerned with the problem of providing rough estimates of the coefficients $t(0), t(1), \ldots$ in the following model for the relation between two zero-mean stationary time series X_1 and X_2:

$$X_2(t) = \sum_{j=0}^{\infty} t(j)X_1(t-j) + N(t),$$

where $\{N(t)\}$ is a zero-mean stationary process, uncorrelated with the "input" process X_1.(See *BD Section 13.1* for more details.)

Before using this program it is necessary to have filed the residual series R_1 obtained from *PEST* after fitting an ARMA model to the series X_1. The residual series R_2, obtained by applying the same ARMA filter to the series X_2, is also needed. This is obtained by applying Option 0 of the Estimation Menu in *PEST* to the data X_2 with the same ARMA model which was fitted to the series X_1. The residuals so obtained constitute the required series R_2.

When Option 2 is selected from the Main Menu of *TRANS* you will be asked for the names of the files containing the "input residuals", R_1, and the "output residuals", R_2. You will then be asked for the order of the moving average relating X_2 to X_1. If you specify the order as $m(< 31)$, estimates will be printed on the screen of the coefficients in the relation,

$$X_2(t) = \sum_{j=0}^{m} t(j)X_1(t-j) + N(t).$$

You may wish to print the estimated coefficients $t(j)$ for later use.

To check which of the estimated coefficients are significantly different from zero and to check the appropriateness of the model, we next plot the sample cross correlations of $R_2(t+h)$ and $R_1(t)$ for $h = -30, -29, \ldots, 30$. These correlations $\hat{\rho}(h)$ are directly proportional to the estimates of $t(h)$ (see *BD Section 13.1*). Sample correlations which fall outside the plotted bounds $(\pm 1.96/\sqrt{n})$ are significantly different from zero (with significance level approximately .05). The plotted values $\hat{\rho}(h)$ should therefore lie within the bounds for $h < b$, where b, the smallest non-negative integer such that $|\hat{\rho}(b)| > 1.96/\sqrt{n}$, is our estimate of the delay parameter. Having identified the delay parameter b, the model previously printed on the screen is revised by setting $t(j) = 0$, $j < b$, giving

$$X_2(t) = \sum_{j=b}^{m} t(j)X_1(t-j) + N(t).$$

After inspecting the graph and recording the estimated delay parameter b and coefficients $t(b), \ldots, t(m)$, press any key and you will be returned to the Main Menu.

EXAMPLE: We shall illustrate the use of Option 2 with reference to the data sets $Y_1 = $DATA\LEAD and $Y_2 = $DATA\SALES on Disk 2.

Analysis of these data sets using *PEST* suggests that differencing at lag 1 and subtracting the means from each of the resulting two series gives rise to series X_1 and X_2 which can be well modelled as zero mean stationary series. The values of the two series are

$$X_1(t) = Y_1(t) - Y_1(t-1) - .0228, \quad t = 2, \ldots, 150,$$

$$X_2(t) = Y_2(t) - Y_2(t-1) - .420, \quad t = 2, \ldots, 150,$$

and the ARMA model fitted by *PEST* to X_1 is

$$X_1(t) = Z(t) - .474Z(t-1), \quad \{Z(t)\} \sim \mathrm{WN}(0, .0779).$$

The residuals R_1 computed from *PEST* have already been filed on Disk 2 under the file name DATA\LRES. Likewise the residuals R_2 obtained by applying the filter $(1 - .474B)^{-1}$ to the series X_2 have been filed on Disk 2 as DATA\SRES. (To generate the latter from *PEST*, input the data set Y_2, difference at lag 1, subtract the mean, input the MA(1) model $X(t) = Z(t) - .474Z(t-1)$, and use Option 0 of the Estimation Menu to compute and file the residuals.)

To find a preliminary transfer function model relating X_2 to X_1, we type in the following entries, starting from the point where the Main Menu of *TRANS* is displayed upon the screen:

```
Order of MA required, m (<31) : 10
PRELIMINARY TRANSFER COEFFICIENTS:
 t( 0) =             .51802010
 t( 1) =             .66472580
 t( 2) =             .33665350
 t( 3) =            4.86250500
 t( 4) =            3.38969400
 t( 5) =            2.60583300
 t( 6) =            2.00288400
 t( 7) =            2.03665600
 t( 8) =            1.52890200
 t( 9) =            1.32632300
 t(10) =             .78603170

MODEL: X2(j)= t(0)X1(j)+...+t(10)X1(j-10)+ N(t)
        <Press any key to continue>
```

FIGURE 5.3. *The estimated coefficients in the transfer function model relating* X_2 *to* X_1

$$2 \hookleftarrow \text{DATA\textbackslash LRES} \hookleftarrow \text{DATA\textbackslash SRES} \hookleftarrow \hookleftarrow 10 \hookleftarrow$$

At this point you will see the screen display in Figure 5.3, showing the estimated coefficients $t(0), t(1), \ldots, t(10)$.

On pressing $\hookleftarrow \hookleftarrow$, you will then see the sample cross correlations shown in Figure 5.4. It is clear from the graph that the correlations are negligible for lags $h < 3$ and that the estimated delay parameter is $b = 3$.

The preliminary model (of order 10) is therefore,

$$X_2(t) = t(3)X_1(t-3) + \cdots + t(10)X_1(t-10) + N(t),$$

where $t(3), \ldots, t(10)$ are as shown in Figure 5.3.

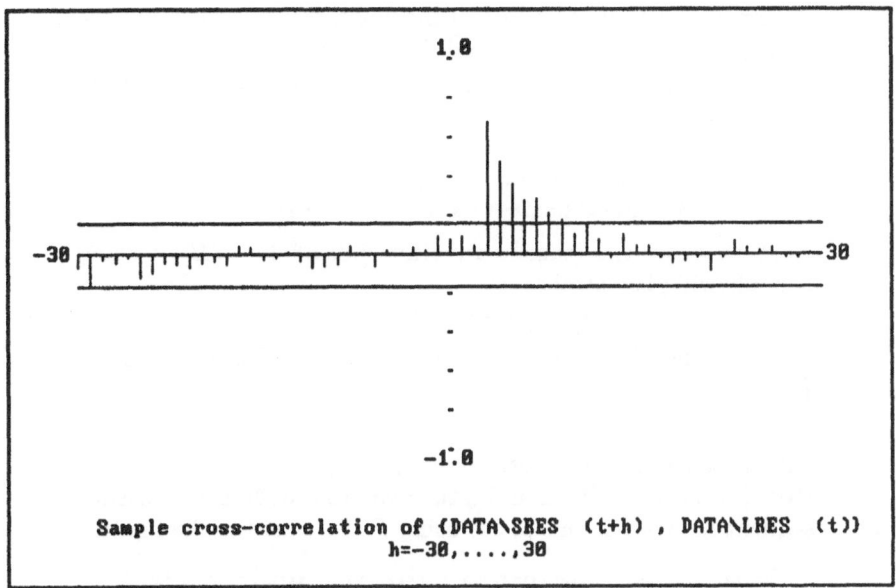

FIGURE 5.4. *The cross correlations of DATA\LRES and DATA\SRES*

5.5   Calculating Residuals from a Transfer Function Model

Option 3 of *TRANS* uses observed values of $X_1(t)$ and $X_2(t)$ and a postulated transfer function model,

$$X_2(t) = B^b(w(0)+w(1)B+\cdots+w(r)B^r)(1-v(1)B-\cdots-v(s)B^s)^{-1}X_1(t)$$
$$+N(t),$$

to generate estimated values $\hat{N}(t)$, $t > m = \max(r+b,s)$, of $N(t)$. The estimates are evaluated from the preceding equation by setting $N(t) = 0$ for $t \le m$ and solving for $N(t), t > m$.

> EXAMPLE: Continuing with the example of Section 5.4, we observe that the estimated moving average transfer function model relating X_2 to X_1 can be well approximated by a model with fewer coefficients, namely,
>
> $$X_2(t) = 4.86B^3(1 - .7B)^{-1}X_1(t) + N(t).$$

To generate estimated values of the noise, $N(t)$, $3 < t \leq 149$, we first generate the series X_1 and X_2 by appropriate differencing and mean correcting of the input series, DATA\LEAD, and the output series, DATA\SALES. This is achieved by typing in the following entries, starting from the point where the Main Menu of *TRANS* is displayed upon the screen:

$$3 \hookleftarrow \text{DATA\textbackslash LEAD} \hookleftarrow \text{DATA\textbackslash SALES} \hookleftarrow 1 \hookleftarrow 1 \hookleftarrow$$

Next we enter the transfer function $4.86B^3(1 - .7B)^{-1}$ using the following key strokes:

$$\hookleftarrow \hookleftarrow 3 \hookleftarrow 0 \hookleftarrow 4.86 \hookleftarrow 1 \hookleftarrow .7 \hookleftarrow$$

You will then be asked for a file name under which to store $\{\hat{N}(t)\}$. The entries

$$\text{NOISE} \hookleftarrow \hookleftarrow$$

will cause the 146 noise estimates, $\{\hat{N}(t), t = 4, \ldots, 149\}$, to be stored in the file NOISE and return you to the Main Menu. Subsequent analysis of this series using *PEST* suggests the model

$$N(t) = (1 - .582B)W(t), \quad \{W(t)\} \sim \text{WN}(0, .0486),$$

for the noise in the transfer function model.

5.6 LS Estimation and Prediction with Transfer Function Models

Option 4 requires specification of a previously fitted ARMA model for the input process and a tentatively specified transfer function (including a model for the noise $\{N(t)\}$). It then estimates the parameters in the model by least squares. The exact Gaussian likelihood is computed using a Kalman filter representation of the model, so that different models can be compared on the basis of their AICC statistics. The Kalman filter representation is also used to give exact best linear predictors of the output series using the fitted model. The mean squared errors of the predictors are estimated using a large-sample approximation for the k-step mean squared error.

The first step is to read in the input and output series and to generate the stationary zero mean series X_1 and X_2 by performing up to two differencing operations followed by mean correction.

The next step is to specify the ARMA model fitted to the series X_1 using *PEST* and to specify the delay parameter, b, the orders, r, s, q and p and preliminary estimates of the coefficients in the transfer function model (*BD Section 13.1*),

$$X_2(t) = \frac{B^b(w(0) + w(1)B + \cdots + w(r)B^r)}{1 - v(1)B - \cdots - v(s)B^s} X_1(t)$$

$$+\frac{1+\theta(1)B+\cdots+\theta(q)B^q}{1-\phi(1)B-\cdots-\phi(p)B^p}W(t).$$

When the model has been specified, the following menu will appear:

```
ESTIMATION AND PREDICTION MENU :
    0. File the current model.
    1. Find least squares estimators.
    2. AICC value and prediction.
    3. File residuals and plot cross-correlations.
       Access to the input residuals filed by PEST
       is required for the latter.
    4. Try a new model.
    5. Enter a new data set.
    6. Return to main menu.
  CHOOSE A NUMBER :
```

Option 1 computes least squares estimators of all the parameters in the model and prints out the parameters of the fitted model. Optimization is typically done with gradually decreasing step-sizes, e.g. .1 for the first optimization, then .01 when the first optimization is complete, and .001 or .0001 for the final optimization.

Once the parameters in the model have been estimated, AICC calculation (for comparison of alternative models) and prediction of future values of the output series can both be done using Option 2. Estimated mean squared errors for the predictors are obtained from large-sample approximations to the k-step prediction errors for the fitted model (see *BD Section 13.1*).

To check the goodness of fit of the model, the residuals $\{\hat{W}(t)\}$ should be examined to check that they resemble white noise and that they are uncorrelated with the residuals from the model fitted to the input process. Option 3 allows them to be filed for further study and checks the cross correlations with the input residuals, provided the latter have been stored in a file which is currently accessible.

> EXAMPLE: Continuing with the example of Section 5.4, we note that the tentative transfer function model we have found relating X_2 to X_1 can now be expressed as
>
> $$X_2(t) = 4.86B^3(1-.7B)^{-1}X_1(t) + (1-.582B)W(t),$$
>
> $$\{W(t)\} \sim \text{WN}(0, .0486),$$
>
> where
>
> $$X_1(t) = (1-.474B)Z(t), \quad \{Z(t)\} \sim \text{WN}(0, .0779).$$

Starting from the screen display of the Main Menu, we first select Option 4 and generate the series X_1 and X_2 by appropriate differencing and mean correcting of the input series,

```
   v(1) =       .70000000

   th(1) =     -.58200000

INPUT AND OUTPUT WN VARIANCES
.77900000E-01  .63645660E-01

INPUT MA COEFFS
 -4.740000E-01

<Press any key to continue>

ESTIMATION AND PREDICTION MENU :
   0. File the current model.
   1. Find least squares estimators.
   2. AICC value and prediction.
   3. File residuals and plot cross-correlations.
      Access to the input residuals filed by PEST
      is required for the latter.
   4. Try a new model.
   5. Enter a new data set.
   6. Return to main menu.
CHOOSE A NUMBER :
```

FIGURE 5.5. *The Estimation and Prediction Menu*

DATA\LEAD, and the output series, DATA\SALES. This is
done by typing:

$$4\hookleftarrow \hookleftarrow \text{DATA}\backslash\text{LEAD}\hookleftarrow \text{DATA}\backslash\text{SALES}\hookleftarrow 1\hookleftarrow 1\hookleftarrow$$

The model previously fitted to X_1 and the orders and coefficients of the tentative transfer function model found in Section
5.4 are now entered as follows:

$$\hookleftarrow \hookleftarrow 0\hookleftarrow 0\hookleftarrow 1\hookleftarrow -.474\hookleftarrow .0779\hookleftarrow 3\hookleftarrow$$
$$0\hookleftarrow 4.86\hookleftarrow 1\hookleftarrow .7\hookleftarrow 1\hookleftarrow -.582\hookleftarrow 0\hookleftarrow \hookleftarrow$$

The specified model will then be displayed on the screen. Press
any key to see the Estimation and Prediction Menu shown in
Figure 5.5.

To obtain least squares estimates of the transfer function coefficients, select Option 1 with step-size .1 by typing

$$1\hookleftarrow .1\hookleftarrow$$

There will be a short delay while optimization is performed. The
screen will then display the new fitted coefficients and white
noise variance, as shown in Figure 5.6.

```
CURRENT MODEL PARAMETERS ARE:
   b    =    3

  w(0)  =    4.91000000

  v(1)  =     .70000000

  th(1) =   -.43200000

INPUT AND OUTPUT WN VARIANCES
.77900000E-01   .59163400E-01

INPUT MA COEFFS
-4.740000E-01

<Press any key to continue>
```

FIGURE 5.6. *The fitted model after using least squares with step-size .1*

To refine the estimates, optimize again with step-size .01 by typing ↩ 1↩ .01↩ and again with step-size .001 by typing ↩ 1↩ .001↩ . The resulting fitted model is shown in Figure 5.7.

To predict 10 values of the original output series DATA\SALES with the fitted model, use Option 2 of the Estimation and Prediction Menu. After typing ↩ 2↩ , you will see the following warning on the screen:

```
Some mathematics coprocessors will have underflow
problems in this option.  If this occurs you will
need to exit from TRANS,switch off the coprocessor
and rerun this option.   The DOS command required
to switch off the coprocessor is
   SET no87=COPROCESSOR OFF
To switch it on again use the command
   SET no87=
```

```
CURRENT MODEL PARAMETERS ARE:
   b    =     3

   w(0)  =      4.71899700

   v(1)  =       .72449990

   th(1) =      -.58249980
INPUT AND OUTPUT WN VARIANCES
.77900000E-01  .48644480E-01

INPUT MA COEFFS
 -4.740000E-01

<Press any key to continue>
```

FIGURE 5.7. *The fitted model after two further optimizations with step-sizes .01 and .001*

> If you have not already filed the current model, it
> may save time to do so now.
>
> Do you wish to file the model (y/n)?

If this warning is applicable to your mathematics coprocessor, you must turn if off as described in the above message. Assuming that this is not necessary, continue by typing n 10↩ . After a short delay you will see the message

> AICC value = .277041E+02

Typing ↩ gives ten predicted values of DATA\SALES, together with the estimated root mean squared errors. The mean squared errors are computed from the large sample approximations described in *BD Section 13.1*. Type y ↩ and the original output series will be plotted on the screen. Then press any key

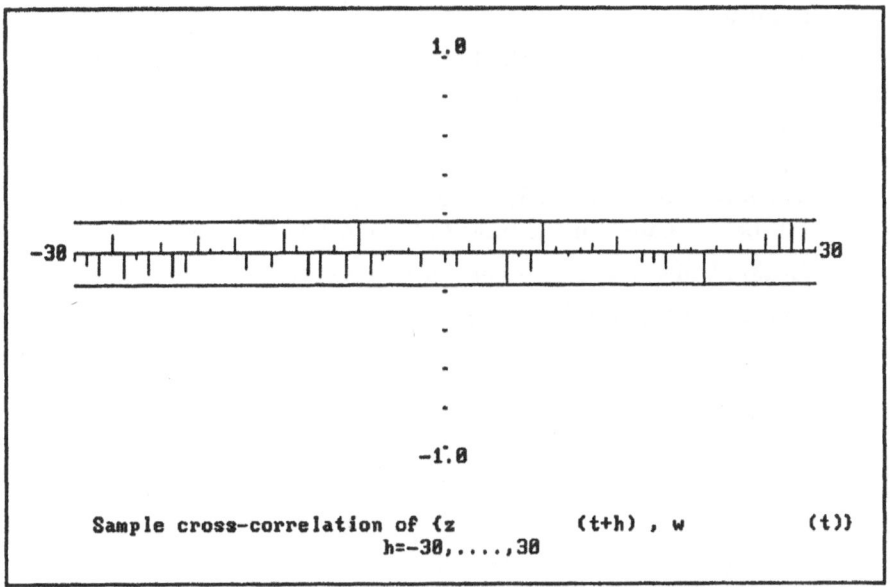

FIGURE 5.8. *The sample cross correlations of the residual series w and z*

and the predictors will also be plotted on the same graph. Type
↩ 0↩ n ↩ to return to the Estimation and Prediction Menu.

To check the goodness of fit of the model, Option 3 of this menu
allows you to file the estimated residuals $\hat{W}(t)$ from the transfer
function model and to check for zero cross-correlations with the
input residuals R_1. To do this type

3↩ w↩ y DATA\LRES↩ z↩ ↩

At this point the estimated residuals, $\hat{W}(t), 3 < t \leq 149$, will
have been stored under the filename w and the corresponding
146 values of $R_1(t)$ under the filename z. You will see on the
screen the sample cross-correlations of these two sets of residuals. For a good fit, approximately 95% of the plotted values
should lie within the plotted bounds. Inspection of the graph
shown in Figure 5.8 indicates that the fitted model is satisfactory from the point of view of residual cross correlations. (The
sample autocorrelations of the residuals filed in w and z are also
found, using *PEST*, to be consistent with those of white noise

sequences.)

After inspecting the graph of sample cross correlations, type ↩
↩ and you will be returned to the Estimation and Prediction
Menu.

Option 4 allows you to input a different preliminary model, for
which the preceding analysis can be repeated. Different models
can be compared on the basis of their AICC statistics.

Option 5 allows you to input a new data set.

Option 6 returns you to the Main Menu.

6

ARVEC

6.1 Introduction

The program *ARVEC* fits a multivariate autoregression of any specified order $p < 21$ to a multivariate time series $\{\mathbf{Y}_t = (Y_{t1}, \ldots, Y_{tm})', t = 1, \ldots, n\}$. To run the program *ARVEC*, type ARVEC $\hookleftarrow$ and enter the appropriate graphics code number. After reading the brief introductory statement, follow the program prompts, entering the dimension $m \leq 6$ of $\mathbf{Y}_t$ and the name of the file containing the observations $\{\mathbf{Y}_t, t = 1, \ldots, n\}$. The data is assumed to be stored as an ASCII file such that row t contains the m components, $\mathbf{Y}_t = (Y_{t1}, \ldots, Y_{tm})'$, each separated by at least one blank space. (The sample size n can be at most 1000.) The value of n will then be printed on the screen and you will be asked if you wish to apply (up to two) differencing transformations to the data.

Examination of the graphs of the component series and their autocorrelations, the latter of which can be checked in *PEST*, indicates whether differencing transformations should be applied to the series $\{\mathbf{Y}_t\}$ before attempting to fit an autoregressive model. (Graphs of the data and the differenced data can also be plotted using *ARVEC* to check for stationarity.) Enter the number of differencing transformations required $(0, 1, 2)$ and the corresponding lags. If, for example, you request two differencing operations with LAG(1)=1 and LAG(2)=12, then the series $\{\mathbf{Y}_t\}$ will be transformed to the differenced series, $(1 - B)(1 - B^{12})\mathbf{Y}_t = \mathbf{Y}_t - \mathbf{Y}_{t-1} - \mathbf{Y}_{t-12} + \mathbf{Y}_{t-13}$. The resulting series is then automatically mean-corrected to generate the series $\{\mathbf{X}_t\}$. The order of the autoregression to be fitted to $\{\mathbf{X}_t\}$ is specified next.

6.1.1 MULTIVARIATE AUTOREGRESSION (*BD Sections 11.3–11.5*)

An m-variate time series $\{\mathbf{X}_t\}$ is said to be a (causal) *multivariate* AR(p) process if it satisfies the recursions

$$\mathbf{X}_t = \Phi_{p1}\mathbf{X}_{t-1} + \cdots + \Phi_{pp}\mathbf{X}_{t-p} + \mathbf{Z}_t, \qquad \{\mathbf{Z}_t\} \sim \text{WN}(0, V_p),$$

where $\Phi_{p1}, \ldots, \hat{\Phi}_{pp}$ are $m \times m$ coefficient matrices, V_p is the error covariance matrix, and $\det(I - z\Phi_{p1} - \cdots - z^p\Phi_{pp}) \neq 0$ for all $|z| \leq 1$. (The p in the subscript Φ_{pj} represents the order of the autoregression.) The coefficient matrices and the error covariance matrix satisfy the multivariate Yule-

Walker equations given by

$$\sum_{j=1}^{p} \Phi_{pj}\Gamma(i-j) = \Gamma(i), \qquad i = 1,\dots,p$$
$$\Gamma(0) - \sum_{j=1}^{p} \Phi_{pj}\Gamma(-j) = V_p.$$

The multivariate version of the Durbin-Levinson algorithm is used to compute the Φ_{pj}'s and V_p recursively. Given observations $\mathbf{x}_1,\dots,\mathbf{x}_n$ of a zero-mean stationary m-variate time series, the fitted AR(p) process $(p < n)$ is then given by

$$\mathbf{X}_t = \hat{\Phi}_{p1}\mathbf{X}_{t-1} + \cdots + \hat{\Phi}_{pp}\mathbf{X}_{t-p} + \mathbf{Z}_t, \qquad \{\mathbf{Z}_t\} \sim \text{WN}(0, \hat{V}_p),$$

where $\hat{\Phi}_{p1},\dots,\hat{\Phi}_{pp}$ and $\hat{V}_p$ are solutions to the Yule-Walker equations above with $\Gamma(h)$ replaced by the sample covariance matrix $\hat{\Gamma}(h), h = 0, 1,\dots,p$. The coefficient estimates are computed using the Durbin-Levinson algorithm.

> EXAMPLE: Consider fitting a multivariate AR(p) process to the bivariate series consisting of the differenced leading indicator (first component) and the differenced sales (second component) time series. The leading indicator-sales data, $\{(Y_{t1}, Y_{t2})', t = 1,\dots,150\}$ are contained in the ASCII file DATA\LS.2. Assuming you have an EGA card, type ARVEC↩ 3↩ 0↩ and you will see the introductory description of the program on the screen. After typing
>
> ↩ 2↩ DATA\LS.2↩
>
> you will be asked if you wish to plot the data. After viewing the graphs of the component series, you will be asked the question,
>
> HOW MANY DIFFERENCING OPERATIONS REQUIRED (<3)?:
>
> Inspection of the autocorrelations and the graphs of the leading indicator and sales time series suggests that both series should be differenced at lag 1 to generate data which are more compatible with realizations from a stationary process. To apply the differencing operator $1 - B$ to $\{\mathbf{Y}_t\}$, type 1 ↩ 1 ↩ . The program then computes the mean-corrected series,
>
> $$\begin{bmatrix} X_{t1} \\ X_{t2} \end{bmatrix} = \begin{bmatrix} Y_{t1} - Y_{t-1,1} \\ Y_{t2} - Y_{t-1,2} \end{bmatrix} - \begin{bmatrix} .02275 \\ .42013 \end{bmatrix}$$
>
> for $t = 2,\dots,150$. At this stage, you have the opportunity to plot the differenced and mean-corrected series to check for any obvious deviations from stationarity (after which you can also change the differencing operations if necessary). In this example, type n in response to the question
>
> Try new differencing operations (y/n)?

since the single differencing at lag 1 appears to be satisfactory. You will then be requested to specify the order $p < \min(21, n)$ of the multivariate AR process to be fitted to $\{\mathbf{X}_t\}$. Try fitting an AR(2) model by typing 2 ↩ . The screen will then display the estimated coefficient matrices $\hat{\mathbf{\Phi}}_{21}, \hat{\mathbf{\Phi}}_{22}$, the estimated white noise covariance matrix $\hat{V}_2$ and the value of the AICC statistic in the following format:

```
PHI( 1)
    -.5096      .0265
    -.7227      .2809

PHI( 2)
    -.1511     -.0103
   -2.1476      .2045

WHITE NOISE COVARIANCE MATRIX, V
    .0765     -.0222
    -.0222    1.4247

AICC =            534.21140
```

The choice $p = 0$ will result in a white noise fit to the data. A useful choice is $p = -1$, which causes the program to find the model with the smallest AICC value (see Section 6.2). To return to the point at which a new value of p may be entered, type n n y.

6.2 Model Selection with the AICC Criterion (*BD Section 11.5*)

The Akaike information criterion (AIC) is a commonly used criterion for choosing the order of a model. This criterion prevents overfitting of a model by effectively assigning a cost to the introduction of each additional parameter. For an m-variate AR(p) process the AICC statistic (a bias-corrected modification of the AIC) computed by the program is

$$\text{AICC} = -2 \ln L(\hat{\mathbf{\Phi}}_{p1}, \ldots, \hat{\mathbf{\Phi}}_{pp}, \hat{V}_p) + 2(pm^2 + 1)n/(n - pm^2 - 2),$$

where L is the Gaussian likelihood of the model based on the n observations, and $\hat{\mathbf{\Phi}}_{p1}, \ldots, \hat{\mathbf{\Phi}}_{pp}, \hat{V}_p$ are the Yule-Walker estimates described in Section 6.1. The order p of the model is chosen to minimize the AICC statistic.

EXAMPLE: For the leading indicator–sales data, the optimal order is found by typing -1 ↩ when asked for the order

```
INDEX OF ORIG.COMPONENT TO BE FORECAST (0<i<3; i=0 to escape):2
FORECASTS :
       TIME       ORIG. Y2        SQRT(MSE)
       151       262.9032         .3084
       152       264.1397         .4254
       153       263.3589         .5640
       154       263.6302        1.4596
       155       263.8901        2.1869
       156       264.1888        2.8736
       157       264.3761        3.5394
       158       264.7391        4.2362
       159       265.0141        4.8996
       160       265.4446        5.5476
<Press any key to continue>
```

FIGURE 6.1. *Forecasts of the next 10 sales values*

of the model. For this example the optimal order is 5 with AICC=114.94230. Note that since the upper right component of each of the coefficient estimates is near 0, $\{X_{t1}\}$ may be modelled separately from $\{X_{t2}\}$. Also note that the first large component in the bottom left corner of the coefficient matrices occurs at lag 3. This suggests that $\{X_{t2}\}$ lags 3 time units behind $\{X_{t1}\}$ (see *BD Example 11.5.1*).

6.3   Forecasting with the Fitted Model (*BD Sections 11.4, 11.5*)

After the fitted model is displayed, the entries y 10↩ will produce forecasts of the next 10 values of $\mathbf{X}_t$. To examine the forecasts and the corresponding standard errors (SQRT(MSE)) of a given component of the series $\{\mathbf{X}_t\}$ or $\{\mathbf{Y}_t\}$ proceed as in the following example.

EXAMPLE: Assuming that the AR(5) model obtained above is still displayed on the screen, the forecasts of the next 10 sales periods can be found by typing y 10 ↩ 0 ↩ 2 ↩ (see Figure 6.1). The forecast of sales at time 153 is 263.3589 with a standard error of .5640. Approximate 95% prediction bounds based on the fitted AR(5) model and assuming that the noise is Gaussian are therefore

$$263.3589 \pm (1.96)(.564).$$

To plot the sales data and the 10 predictors, type ↩ y ↩ ↩ To get the forecasts of the leading indicator series 10 steps ahead, press any key and type ↩ 0↩ 1 ↩ .

After escaping from the forecasting part of *ARVEC* (type 0 ↩), you have the options to file the one-step prediction errors for $\{\mathbf{X}_t\}$,

$$\mathbf{X}_t - \hat{\mathbf{\Phi}}_{p1}\mathbf{X}_{t-1} - \cdots - \hat{\mathbf{\Phi}}_{pp}\mathbf{X}_{t-p}, \qquad t = p+1,\ldots,n$$

and to fit a different model (i.e., one with a different value of p) to the series $\{\mathbf{X}_t\}$.

7

ARAR

7.1 Introduction

To run the program *ARAR*, type ARAR↩ . After entering the appropriate graphics code number you will see a brief introductory statement. The program is an adaptation of the ARARMA forecasting scheme of Newton and Parzen (see *The Accuracy of Major Forecasting Procedures*, ed. Makridakis et al., John Wiley, 1984, pp.267 - 287). The latter was found to perform extremely well in the forecasting competition of Makridakis, the results of which are described in the book. The ARARMA scheme has a further advantage over most standard forecasting techniques in being more readily automated.

On typing ↩ you will be given the options of entering a data set or exiting from the program. Once you have entered a data set and pressed ↩ you will see the following Main Menu printed on the screen:

```
MAIN MENU:
    1. Input a new data set.
    2. Plot the data.
    3. Determine the memory-shortening polynomial and
       fit a subset AR model to the transformed data.
    4. Bypass memory-shortening and fit a subset AR
       model to the original data.
    5. Exit from program.
    CHOOSE A NUMBER:
```

7.1.1 MEMORY SHORTENING

Given a data set $\{Y_t, t = 1, 2, \ldots, n\}$, the first step is to decide whether or not the process is "long-memory", and if so to apply a memory-shortening transformation before attempting to fit an autoregressive model. The differencing operations permitted by *PEST* are examples of memory-shortening transformations, however the ones allowed by *ARAR* are more general. There are two types allowed:

$$\tilde{Y}_t = Y_t - \hat{\phi}(\hat{\tau})Y_{t-\hat{\tau}}, \tag{1}$$

and

$$\tilde{Y}_t = Y_t - \hat{\phi}_1 Y_{t-1} - \hat{\phi}_2 Y_{t-2}. \tag{2}$$

With the aid of the five-step algorithm described below, we shall classify $\{Y_t\}$ and take one of the following three courses of action.

- **L.** Declare $\{Y_t\}$ to be long-memory and form $\{\tilde{Y}_t\}$ using (1).

- **M.** Declare $\{Y_t\}$ to be moderately long-memory and form $\{\tilde{Y}_t\}$ using (2).

- **S.** Declare $\{Y_t\}$ to be short-memory.

If the alternatives L or M are chosen then the transformed series $\{\tilde{Y}_t\}$ is again checked. If it is found to be long-memory or moderately long-memory, then a further transformation is performed. The process continues until the transformed series is classified as short-memory. The program *ARAR* allows at most three memory-shortening transformations. It is very rare to require more than two. The algorithm for deciding between L, M and S can be described as follows:

1. For each $\tau = 1, 2, \ldots, 15$, we find the value $\hat{\phi}(\tau)$ of ϕ which minimizes

$$\text{Err}(\phi, \tau) = \frac{\sum_{t=\tau+1}^{n} [Y_t - \phi Y_{t-\tau}]^2}{\sum_{t=\tau+1}^{n} Y_t^2}.$$

We then define

$$\text{Err}(\tau) = \frac{\sum_{t=\tau+1}^{n} [Y_t - \hat{\phi}(\tau) Y_{t-\tau}]^2}{\sum_{t=\tau+1}^{n} Y_t^2},$$

and choose the lag $\hat{\tau}$ to be the value of τ which minimizes $\text{Err}(\tau)$.

2. If $\text{Err}(\hat{\tau}) \leq 8/n$, go to L.

3. If $\hat{\phi}(\hat{\tau}) \geq .93$ and $\hat{\tau} > 2$, go to L.

4. If $\hat{\phi}(\hat{\tau}) \geq .93$ and $\hat{\tau} = 1$ or 2, determine the values $\hat{\phi}_1$ and $\hat{\phi}_2$ of ϕ_1 and ϕ_2 which minimize

$$\sum_{t=3}^{n} [Y_t - \phi_1 Y_{t-1} - \phi_2 Y_{t-2}]^2.$$

Go to M.

5. If $\hat{\phi}(\hat{\tau}) < .93$, go to S.

7.1.2 FITTING A SUBSET AUTOREGRESSION

Let $\{S_t, t = 1, \ldots, T\}$ denote the memory-shortened series derived from $\{Y_t\}$ by the algorithm of the previous section and let $\overline{S}$ denote the sample mean of $S_1, \ldots, S_T$.

The next step in the modelling procedure is to fit an autoregressive process to the mean-corrected series,

$$X_t = S_t - \overline{S}, \quad t = 1, \ldots, T.$$

The fitted model has the form

$$X_t = \phi_1 X_{t-1} + \phi_{l_1} X_{t-l_1} + \phi_{l_2} X_{t-l_2} + \phi_{l_3} X_{t-l_3}$$

$$+ Z_t,$$

where $\{Z_t\} \sim \mathrm{WN}(0, \sigma^2)$, and, for given lags, l_1, l_2, and l_3, the coefficients ϕ_j and the white noise variance σ^2 are found from the Yule-Walker equations,

$$\begin{bmatrix} 1 & \hat{\rho}(l_1 - 1) & \hat{\rho}(l_2 - 1) & \hat{\rho}(l_3 - 1) \\ \hat{\rho}(l_1 - 1) & 1 & \hat{\rho}(l_2 - l_1) & \hat{\rho}(l_3 - l_1) \\ \hat{\rho}(l_2 - 1) & \hat{\rho}(l_2 - l_1) & 1 & \hat{\rho}(l_3 - l_2) \\ \hat{\rho}(l_3 - 1) & \hat{\rho}(l_3 - l_1) & \hat{\rho}(l_3 - l_2) & 1 \end{bmatrix} \begin{bmatrix} \phi_1 \\ \phi_{l_1} \\ \phi_{l_2} \\ \phi_{l_3} \end{bmatrix}$$

$$= \begin{bmatrix} \hat{\rho}(1) \\ \hat{\rho}(l_1) \\ \hat{\rho}(l_2) \\ \hat{\rho}(l_3) \end{bmatrix},$$

and

$$\sigma^2 = \hat{\gamma}(0)[1 - \phi_{l_1}\hat{\rho}(1) - \phi_{l_1}\hat{\rho}(l_1) - \phi_{l_2}\hat{\rho}(l_2) - \phi_{l_3}\hat{\rho}(l_3)],$$

where $\hat{\gamma}(j)$ and $\hat{\rho}(j), j = 0, 1, 2, \ldots$, are the sample autocovariances and autocorrelations of the series $\{X_t\}$.

The program computes the coefficients ϕ_j for each set of lags such that

$$1 < l_1 < l_2 < l_3 \leq m$$

where m can be chosen to be either 13 or 26. It then selects the model for which the Yule-Walker estimate σ^2 is minimum and prints out the lags, coefficients and white noise variance for the fitted model.

A slower procedure chooses the lags and coefficients (computed from the Yule-Walker equations as above) which maximize the Gaussian likelihood of the observations. For this option the maximum lag m is 13.

The options are displayed in the following Subset AR Menu which appears on the screen when memory-shortening has been completed (or when you opt to by-pass memory shortening and fit a subset AR to the original (mean-corrected) data).

SUBSET AR MENU :
1. Find the four-coefficient Yule-Walker model with
   minimum WN variance estimate (maximum lag = 13)
2. Find the four-coefficient Yule-Walker model with
   minimum WN variance estimate (maximum lag = 26)
3. Find the four-coefficient Yule-Walker model with
   maximum Gaussian likelihood (maximum lag = 13).
4. Return to main menu.
CHOOSE A NUMBER :

7.2 Running the Program

To determine an ARAR model for the given data set $\{Y_t\}$ and to use it to forecast future values of the series, we first input the data set. Following the appearance on the screen of the Main Menu, we then type 3↩ to find the best memory-shortening filter. After a short time delay the coefficients $1, \psi_1, \ldots, \psi_k$ of the chosen filter will be displayed on the screen. The memory shortened series is

$$S_t = Y_t + \psi_1 Y_{t-1} + \cdots + \psi_k Y_{t-k}.$$

Type ↩ and the Subset AR Menu will appear. Option 1 fits an autoregression with four non-zero coefficients to the mean-corrected series $X_t = S_t - \overline{S}$, choosing the lags and coefficients which minimize the Yule-Walker estimate of white noise variance. After a short delay, the optimal lags and corresponding coefficients in the model

$$X_t = \phi_1 X_{t-1} + \phi_{l_1} X_{t-l_1} + \phi_{l_2} X_{t-l_2} + \phi_{l_3} X_{t-l_3}$$

$$+Z_t,$$

will be printed on the screen. The coefficients ξ_j of B^j in the overall whitening filter,

$$\xi(B) = (1 + \psi_1 B + \cdots + \psi_k B^k)(1 - \phi_1 B - \phi_{l_1} B^{l_1} - \phi_{l_2} B^{l_2} - \phi_{l_3} B^{l_3}),$$

are also printed.

Type ↩ and you will be asked for the number of future values of $\{Y_t\}$ to be predicted. Enter the required number and type ↩ ↩ to see the graph of the original data. Type ↩ again and the predicted values will be added to the graph. Type ↩ and you will be asked if you wish to file the predictors. Following this you will be returned to the Subset AR Menu, from which you may either select one of the other fitting options or return to the Main Menu from which you may leave the program.

EXAMPLE: To use the program *ARAR* to predict 24 values of the data file DATA\DEATHS on Disk 2, use the following sequence

```
      < Finding best memory shortening polynomial>
BEST LONG-MEMORY LAG
   12
LAGGED AR COEFFICIENT
   9.778940E-01
RESIDUAL S.S./TOTAL S.S.
   3.668375E-03

BEST LONG-MEMORY LAG
   1
LAGGED AR COEFFICIENT
   7.024037E-01
RESIDUAL S.S./TOTAL S.S.
   4.838039E-01

      < Memory shortening is now complete>

COEFFICIENT OF B**J IN MEMORY-SHORTENING POLYNOMIAL,   J=0,1,... :
   1.0000      .0000      .0000      .0000      .0000
    .0000      .0000      .0000      .0000      .0000
    .0000      .0000     -.9779

   <Press any key to continue>
```

FIGURE 7.1. *Memory-shortening filter selected for DATA\DEATHS*

of entries (for graphics we are assuming that you have an EGA card):

ARAR ↩ 3↩ 0↩ ↩ 1↩ DATA\DEATHS ↩ ↩ 3↩

At this point the screen displays the coefficients of the selected memory-shortening filter. Figure 7.1 shows that the chosen filter is $(1 - .9779B^{12})$.

To continue, type

↩ 1↩

The program then fits a four-coefficient AR model to the mean-corrected memory-shortened data, with maximum lag 13, selecting the model with minimum estimated white-noise variance. The fitted model is displayed in Figure 7.2.

To predict 24 future values of the series, type

↩ 24↩

At this stage the screen will show the square root of the *observed* mean squared error of the one-step predictors of the data itself.

```
MEAN OF SHORT-MEMORY SERIES   =      23.2217
LENGTH OF SHORT-MEMORY SERIES =      60


  SUBSET AR MENU :
  1. Find the four-coefficient Yule-Walker model with
     minimum WN variance estimate (maximum lag = 13).
  2. Find the four-coefficient Yule-Walker model with
     minimum WN variance estimate (maximum lag = 26).
  3. Find the four-coefficient Yule-Walker model with
     maximum Gaussian likelihood  (maximum lag = 13).
  4. Return to main menu.
  CHOOSE A NUMBER : 1
Optimal lags   :            1         3        12        13
Optimal coeffs :        .5915     .2093    -.3022     .2970
WN Variance    :   .12314E+06
COEFFICIENTS OF OVERALL WHITENING FILTER :
  1.0000     -.5915      .0000    -.2093      .0000
   .0000      .0000      .0000     .0000      .0000
   .0000      .0000     -.6757     .2814      .0000
   .2047      .0000      .0000     .0000      .0000
   .0000      .0000      .0000     .0000     -.2955
   .2904

  <Press any key to continue>
```

FIGURE 7.2. *The four-coefficient autoregression fitted to the memory-shortened DATA\DEATHS series*

To plot the predictors of future values type

↩ ↩

and you will see the original data with the 24 predictors plotted on the same graph as in Figure 7.3. To exit from the program type ↩ 0↩ n 4↩ 5↩ .

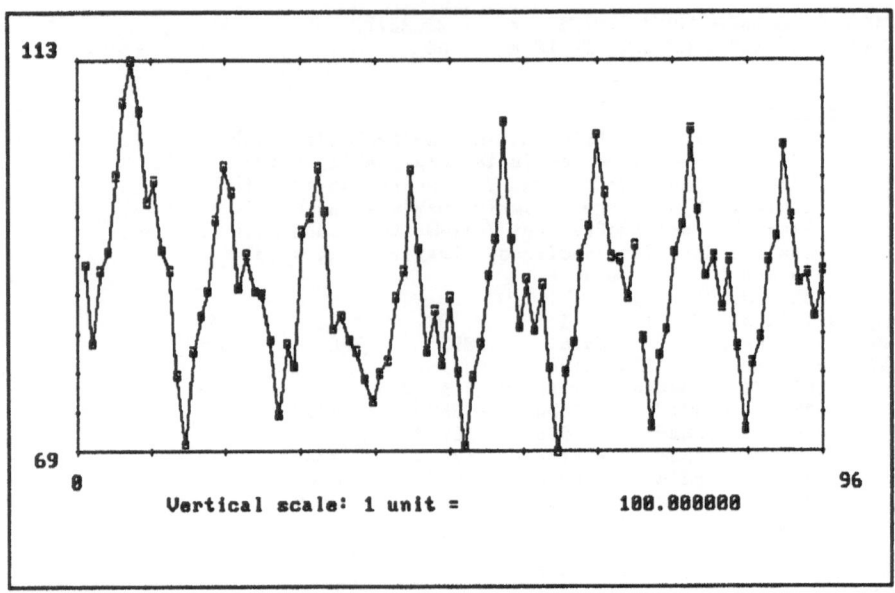

FIGURE 7.3. *The data set DATA\DEATHS with 24 predicted values*

Appendix A

Word6: A Screen Editor

ANTHONY E. BROCKWELL

A.1 Basic Editing

The cursor can be moved with the four cursor control keys. The <End> key moves the cursor to the end of the line, and the <Home> key to the first character of the line. <PgUp> and <PgDn> move the cursor 22 lines up and 22 lines down, respectively. <Ctrl>-<Home> and <Ctrl>-<End> move the cursor to the beginning and end of the text respectively (<Ctrl>-<Home> means holding down the <Ctrl> key while pressing the <Home> key).

The backspace key <←> (upper right of keyboard) deletes the character at the cursor position and moves the cursor back one space. The <Del> key deletes the character at the cursor position without moving the cursor. To merge two lines, move the cursor to the far left of the screen (using <Home> and then the left arrow) and press the <←> key. The line will then be moved up and put on the end of the line above.

The <Ins> key toggles insert and overwrite modes. In insert mode characters will be inserted into the text at the current cursor position. In overwrite mode they replace the old character and the <Enter> key moves the cursor to the next line without inserting a new line. At the bottom of the screen, a message shows whether you are in insert or overwrite mode.

A.2 Alternate Keys

To perform special functions, *WORD6* makes use of the <Alt> key. The <Alt> key works in the same way as the <Shift> key. To enter <Alt>-X, for example, press the <Alt> key, and while still holding it down, press X. Note — either X or x will do, as the computer does not differentiate between upper and lower case alternate keys.

The two most essential <Alt>-keys are <Alt>-R and <Alt>-W. To read an ASCII file into the editor, type <Alt>-R, and then enter the name of the file to be read in. (Alternatively, you can type WORD6 FNAME to begin editing an existing file called FNAME.) When you have finished editing or creating a file, <Alt>-W can be used to write the file to disk. <Alt>-V is the same as <Alt>-R, but it eliminates the high bit of every character read in, so it can be used to read a WordStar file into *WORD6* for editing.

To exit from *WORD6*, enter <Alt>-X. If you have edited a file without saving it, you will be asked whether you really want to exit without saving the file. To save the file, answer n and then use <Alt>-W.

<Alt>-D and <Alt>-I can be used to delete and insert large sections of text quickly. <Alt>-D deletes the entire line at the current cursor position and moves all the text below it up one line. <Alt>-I inserts a blank line above the current cursor position.

If you have a color monitor, <Alt>-Z can be used to change the screen color.

A.3   Printing a File

To print a file, use <Alt>-W as though writing a file. Then when prompted for the file name, enter LPT1 (or possibly LPT2 if you have two printers). This is a DOS filename which allows the printer to be treated as though it were a file.

To make use of special printer control codes (for underlining, bold-face, etc.) enter these codes directly into the document. Use <Alt>-0 to redefine <Alt>-(1-9) by ASCII code, and then any combination of control codes can be sent to the printer.

A.4   Merging Two or More Files

The <Alt>-R command does not replace the old document with a new one. It merges the new file into the current text. If there is no current text — as after using <Alt>-N or just after entering *WORD6* from the DOS prompt — the new file will obviously not be merged. For example, to merge a fifty line file between the tenth and eleventh lines of an old sixty line file, read in the sixty line file, insert a blank line between its tenth and eleventh lines, position the cursor anywhere on the blank line, and then read in the fifty line file using <Alt>-R. To merge the new file onto the end of the old one, just position the cursor at the end of the old one using <Ctrl>-<End>, press ↩ (optional), and read in the new one.

A.5   Margins and Left and Centre Justification

<Alt>-L and <Alt>-P set left and right margins, respectively. The margin will be set at wherever the cursor is when the key is pressed. The shading over the tab settings will change to show only what is included between the left and right margin. Text will automatically wrap around to the left

margin on the next line if the cursor moves past the right margin on the current line. To left-justify text, press <Ctrl>-L (like <Alt>-L, except use the <Ctrl> key instead of the <Alt> key). The current line will be moved so that it starts right on the left margin. <Ctrl>-M will centre-justify text by placing it centrally between the right and left margins.

A.6   Tab Settings

<Alt>-T sets or removes a tab setting. If there is a tab setting at the current cursor position, it will be removed, if there is no tab setting, one will be added. Tab settings are indicated by little white hats at the bottom of the screen. When the tab key is pressed, the cursor will automatically move to the next tab position. Example: To get rid of the next tab setting, press the tab key to move there, and then press <Alt>-T to remove the setting. The hat marking that tab setting will disappear.

A.7   Block Commands

Large sections of text can be moved or erased as follows using the <Alt>-M command. Move to the first line of the section to be marked and press <Alt>-M. Then move to the last line and press <Alt>-M again. The entire block between and including the two lines will change color to show that it has been marked. After marking a block, the <Alt>-E and <Alt>-C commands can be used. <Alt>-E deletes the entire block. <Alt>-C makes a second copy of the block after the line at the current cursor position. For instance, to delete the entire text, press <Ctrl>-<Home>, <Alt>-M, <Ctrl>-<End>, <Alt>-M and then <Alt>-E will erase the entire text.
   Note:

1. Only one block can exist at once. <Alt>-C makes a copy of the old block and leaves it marked.

2. To unmark a block, press <Alt>-M. If a block already exists, <Alt>-M removes the marking.

3. To move a section of text, mark it, move the cursor to the line before the new desired position and press <Alt>-C, and then press <Alt>-E to get rid of the old block.

If you wish to write only part of the text to a file, mark the required block, and then press <Alt>-B. You will be prompted for the file name.
   Vertical blocks may also be manipulated by using <Alt>-F. Mark each end of the block by pressing <Alt>-F. To delete a marked block press <Alt>-G. To move a marked block to the right or left, press <Alt>-U and

use the arrow keys. When the marked block is appropriately located press
↩ . To unmark the block press <Alt>-F again.

A.8   Searching

To locate a certain word or set of characters, use <Alt>-S. You will be
prompted for a string to search for and what to replace it with. If you
want to search and not replace, just press ↩ when asked **Replace with ?.**
You will then be asked the question **Ignore case (Y/N) ?.** (If you answer
N then a search for *The* will not find *the.*) The cursor will then be moved
to the first occurrence of the string after the current cursor position. If
the string is not found in the text, the cursor will reappear at the end of
the file. Searching and replacing is always global, but can be aborted with
the <Esc> key. Each time the string is found, you will be prompted as to
whether or not to replace it. If you enter N or n, the search will go on to the
next occurrence of the string. Since a search always starts at the current
cursor position, it is usually a good idea to go to the beginning of the text
using <Ctrl>-<Home> before carrying out a search.

   Example: To replace every occurrence of *this* in the text with *that*, go to
the beginning of the text by pressing <Ctrl>-<Home>. Then press <Alt>-S.
Then enter **this** and then enter **that**. *WORD6* will then give the prompt
Replace (Y/N)? for every occurrence of this in the text. If you enter y or Y
the *this* at the cursor position will be changed to a *that*.

   <Alt>-Q repeats the last search, and does not replace.

A.9   Special Characters

By making use of <Alt>-(1-9), *WORD6* can access characters which cannot normally be accessed from the keyboard. Each time *WORD6* is run,
a set of some of the more useful Greek letters are loaded into the keys
<Alt>-1, <Alt>-2, ... <Alt>-9. However these can be redefined by ASCII
code by pressing <Alt>-0.

   A box can be created by using ASCII codes 192, 196, 217, 179, 218 and
191. These each display a different segment of the box. Press <Alt>-0 and
enter these six numbers for six of the nine <Alt>-keys. Then by pressing
<Alt>-(1-9), these segments of the box can be put on the screen and
edited to the correct position.

A.10   Function Keys

Function can be defined to be any string of up to forty characters. It can save time to redefine commonly used phrases as function keys (for instance `write(*,*)` in Fortran.) Press `<F10>` to redefine a function key. When asked which one to define, press the function key you wish to assign a string to.[1] Then enter the string. You may define up to nine different function keys at once.

A.11   Editing Information

At the bottom of the screen is a list of parameters. At the far left is a message `<Alt>-h = Help`. Next to that is either Insert or Overwrite. This is the current editing mode, which can be toggled using the `<Ins>` key. Next to that a number displays the column number of the cursor (anywhere from 1 to 65535). At the far right are two numbers, separated by a slash. The number on the left of the slash is the number of the line at which the cursor is currently located. The number on the right of the slash is the total number of lines in the document.

On the line above all this information, a series of hats may be displayed. These are all the tab settings. In addition to the tab settings, this line is shaded to show the left and right margins.

---

[1]The tab key will appear as a small circle when used in the definition of a function key, and will be decoded when the function key is pressed while editing. Thus function key definitions including tabs will be placed on the screen as though the tab key is pressed at the position it appears on the screen. It is not converted into a set number of spaces to be put on the screen.

Appendix B

Data Sets

USPOP U.S. Population at ten-year intervals, 1790–1980 (U.S.Bureau of the Census). *BD Example 1.1.2.*

STRIKES Strikes in the U.S.A., 1951–1980 (Bureau of Labor Statistics). *BD Example 1.1.3.*

SUNSPOTS The Wolfer sunspot numbers, 1770–1869. *BD Example 1.1.5.*

DEATHS Monthly accidental deaths in the U.S.A., 1973–1978 (National Safety Council). *BD Example 1.1.6.*

AIRPASS International airline passenger monthly totals (in thousands), Jan. 49 – Dec. 60. From Box and Jenkins (*Time Series Analysis: Forecasting and Control, 1970*). *BD Example 9.2.2.*

E911 200 simulated values of an ARIMA(1,1,0) process. *BD Example 9.1.1.*

E921 200 simulated values of an AR(2) process. *BD Example 9.2.1.*

E923 200 simulated values of an ARMA(2,1) process. *BD Example 9.2.3.*

E951 200 simulated values of an ARIMA(1,2,1) process. *BD Example 9.5.1.*

E1021 Sinusoid plus simulated Gaussian white noise. *BD Example 10.2.1.*

E1042 160 simulated values of an MA(1) process. *BD Example 10.4.2.*

LEAD Leading Indicator Series from Box and Jenkins (*Time Series Analysis: Forecasting and Control, 1970*). *BD Example 11.2.2.*

SALES Sales Data from Box and Jenkins (*Time Series Analysis: Forecasting and Control, 1970*). *BD Example 11.2.2.*

E1241 200 values of a simulated fractionally integrated MA(1) series. *BD Example 13.2.1.*

E1251 200 values of a simulated MA(1) series with standard Cauchy white noise. *BD Example 13.3.1.*

E1252 200 values of a simulated AR(1) series with standard Cauchy white noise. *BD Example 13.3.2.*

APPA Lake level of Lake Huron in feet (reduced by 570), 1875–1972. *BD Appendix Series A.*

APPB Dow Jones Utilities Index, Aug.28–Dec.18, 1972. *BD Appendix Series B.*

APPC Private Housing Units Started, U.S.A. (monthly). From Makridakis competition, series 922. *BD Appendix Series C.*

APPD Industrial Production, Austria (quarterly). From Makridakis competition, Series 337. *BD Appendix Series D.*

APPE Industrial Production, Spain (monthly). From Makridakis competition, Series 868. *BD Appendix Series E.*

APPF General Index of Industrial Production (monthly). From Makridakis competition, Series 904. *BD Appendix, Series F.*

APPG Annual Canadian Lynx Trappings, 1821–1934. *BD Appendix Series G.*

APPH Annual Mink Trappings, 1848–1911. *BD Appendix Series H.*

APPI Annual Muskrat Trappings, 1848–1911. *BD Appendix Series I.*

APPJ Simulated input series for transfer function model. *BD Appendix Series J.*

APPK Simulated output series for transfer function model. *BD Appendix Series K.*

LRES Whitened Leading Indicator Series obtained by fitting an MA(1) to the mean-corrected differenced series LEAD. *BD Section 13.1.*

SRES Residuals obtained from the mean-corrected differenced SALES data when the filter used for whitening the mean-corrected differenced LEAD series is applied. *BD Section 13.1.*

APPJK.2 The two series APPJ and APPK (see above) in bivariate format for analysis by ARVEC.

LS.2 Lead-Sales data in bivariate format for analysis by ARVEC.

Index

Springer for Statistics

Further Reading on Time Series —

P.J. Brockwell and R.A. Davis,
Colorado State University, Fort Collins, CO
Time Series
Theory and Methods
Second Edition

Further details on time series are discussed in this text which can be used in conjunction with the ITSM programs.

From the reviews of the first edition:
"To get right to the point, I like this book a lot. . . Along with its excellent exposition, there are many features of the book that make it attractive as a textbook or reference . . . well organized and clearly written. . . extremely useful. . . self-contained."
— **Journal of the American Statistical Association**

". . . does a splendid job in presenting theory, including the basic material of Hilbert spaces, recording detailed calculations involving asymptotic approximations, and showing the use of the method for identifying models and estimating parameters on data sets. . . thoroughly recommended."
— **Short Book Reviews**

This new second edition contains a large number of additions and corrections throughout the text, including the incorporation of a new chapter on state-space models.

1991/592 pp., 126 illus./Hardcover/ISBN 0-387-97429-6
Springer Series in Statistics

Ordering Information: *In North America:* Call 1-800-SPRINGER (1-800-777-4643), in NJ call 201-348-4033. Or Write to Springer-Verlag New York, Inc., 175 Fifth Avenue, New York, NY 10010. *Outside North America:* Write to Springer-Verlag GmbH & Co. KG, Tiergartenstrasse 17, D-6900 Heidelberg 1, FRG. (*Attn: Mathematics Promotion*).

Springer-Verlag

New York • Berlin • Heidelberg • Vienna • London • Paris • Tokyo • Hong Kong • Barcelona